EXPERIMENTATION, VALIDATION, AND UNCERTAINTY ANALYSIS FOR ENGINEERS

EXPERIMENTATION, VALIDATION, AND UNCERTAINTY ANALYSIS FOR ENGINEERS

THIRD EDITION

HUGH W. COLEMAN and W. GLENN STEELE

WILEY

A JOHN WILEY & SONS, INC.

Library of Congress Cataloging-in-Publication Data:

Coleman, Hugh W.
 Experimentation, validation, and uncertainty analysis for engineers / Hugh W. Coleman, W. Glenn Steele.. −3rd ed.
  p. cm.
 Rev. ed. of: Experimentation and uncertainty analysis for engineers.
 Includes bibliographical references and index.
 ISBN 978-0-470-16888-2 (cloth)
 1. Engineering–Experiments. 2. Uncertainty. I. Steele, W. Glenn. II. Coleman, Hugh W. Experimentation and uncertainty analysis for engineers. III. Title.
 TA153.C66 2009
 620.0072—dc22

                                                    2009018566

Printed in the United States of America

10 9 8 7 6 5 4 3 2 1

\ 0   0601054 ⌐

This one's for Anne and our four decades plus journey together; and for William Sidney Coleman, who's new to all of this.

H.W.C

For Cherie, my wife and best friend; our grandchildren, Taylor, Ashley, Cameron, Anna, Allison, Ava Grace, and Mary Margaret; and their parents, Amy and Ed, Holly and Scott, and Carrie and Justin

W.G.S.

CONTENTS

PREFACE

The third edition of this book incorporates significant revisions, updates, and new material. Key among these is the new chapter on "Validation of Simulations." The incorporation of Validation in the title of the book signifies the high level of importance that uncertainty analysis is playing in the verification and validation of engineering and scientific computational tools. We were members of the committee that developed the new ASME *Standard for Verification and Validation in Computational Fluid Dynamics and Heat Transfer*, V&V20-2009, and we have presented this methodology in the new Chapter 6 of the book. The application of uncertainty analysis concepts in validation of simulations is also discussed as the subject arises naturally in Chapters 1 through 5.

In this edition, we present two primary methods for uncertainty propagation: the Taylor Series Method (TSM) and the Monte Carlo Method (MCM). The TSM approach was used in the earlier editions of the book and follows the methodology of the international standard, the ISO *Guide to the Expression of Uncertainty in Measurement*. With fast, inexpensive computers, the MCM approach is now a powerful tool for uncertainty propagation. Our treatment of the subject follows the 2008 supplement to the ISO guide from the Joint Committee for Guides in Metrology, *Evaluation of Measurement Data—Supplement 1 to the "Guide to the Expression of Uncertainty in Measurement"—Propagation of Distributions Using a Monte Carlo Method*. The new Chapter 3 in the book gives the basic information and methodology for both methods and provides an example of their use in uncertainty determination.

In this edition, we deal with uncertainties at the standard deviation level, or as standard uncertainties, using the ISO guide nomenclature. We characterize uncertainties both by their effects on the measurements, systematic or random, and by their evaluation method, ISO Type A or B. We also present the methodologies and approximations for expanding the uncertainty to a given percent level of confidence.

Even with the substantial revisions and additions we have made, our objective for this edition is the same as for the first edition: to present a logical approach to experimentation (*and validation*) through the application of uncertainty analysis in the planning, design, construction, debugging, execution, data analysis, and reporting phases of experimental (*and validation*) programs. The style of the book is directed toward its use as a teaching tool or as a reference for practicing engineers and scientists and is based on our many years of experience in presenting the material to junior/senior engineering students, to graduate students from a broad range of disciplines, and in classes of our two-day short course (www.uncertainty-analysis.com) to industry and laboratory technical personnel.

We would like to acknowledge the invaluable contributions of our students to this work and also the contributions and recommendations of colleagues at our universities and on committees and groups with which we have served. It is always stimulating to present the material to classes at the university or at short courses and to see how the students quickly see the myriad of applications of applied uncertainty analysis to their specific tests and analyses.

HUGH W. COLEMAN
University of Alabama in Huntsville
W. GLENN STEELE, JR.
Mississippi State University

March 2009

EXPERIMENTATION, VALIDATION, AND UNCERTAINTY ANALYSIS FOR ENGINEERS

1

EXPERIMENTATION, ERRORS, AND UNCERTAINTY

When the word *experimentation* is encountered, most of us immediately envision someone in a laboratory "taking data." This idea has been fostered over many decades by portrayals in periodicals, television shows, and movies of an engineer or scientist in a white lab coat writing on a clipboard while surrounded by the piping and gauges in a refinery or by an impressive complexity of laboratory glassware. In recent years, the location is often a control room filled with computerized data acquisition equipment with lights blinking on the racks and panels. To some extent, the manner in which laboratory classes are typically implemented in university curricula also reinforces this idea. Students often encounter most instruction in experimentation as demonstration experiments that are already set up when the students walk into the laboratory. Data are often taken under the pressure of time, and much of the interpretation of the data and the reporting of results is spent on trying to rationalize what went wrong and what the results "would have shown if...."

Experimentation is not just data taking. Any engineer or scientist who subscribes to the widely held but erroneous belief that experimentation is making measurements in the laboratory will be a failure as an experimentalist. The actual data-taking portion of a well-run experimental program generally constitutes a small percentage of the total time and effort expended. In this book we examine and discuss the steps and techniques involved in a logical, thorough approach to the subject of experimentation.

1-1 EXPERIMENTATION

1-1.1 Why Is Experimentation Necessary?

Why are experiments necessary? Why do we need to study the subject of experimentation? The experiments run in science and engineering courses demonstrate physical principles and processes, but once these demonstrations are made and their lessons taken to heart, why bother with experiments? With the laws of physics we know, with the sophisticated analytical solution methods we study, with the increasing knowledge of numerical solution techniques, and with the awesome computing power available, is there any longer a need for experimentation in the real world?

These are fair questions to ask. To address them, it is instructive to consider Figure 1.1, which illustrates a typical *analytical* approach to finding a solution to a physical problem. Experimental information is almost always required at one or more stages of the solution process, even when an analytical approach is used. Sometimes experimental results are necessary before realistic assumptions and idealizations can be made so that a mathematical model of the real-world process can be formulated using the basic laws of physics. In addition, experimentally determined information is generally present in the form of physical property values and the auxiliary equations (e.g., equations of state) necessary for obtaining a solution. So we see that even in situations in which the solution approach is analytical (or numerical) information from experiments is included in the solution process.

From a more general perspective, experimentation lies at the very foundations of science and engineering. *Webster's* [1] defines science as "systematized

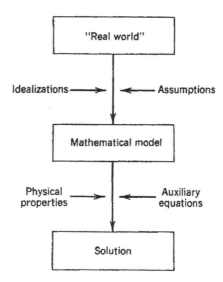

Figure 1.1   Analytical approach to solution of a problem.

knowledge derived from observation, study, and experimentation carried on in order to determine the nature or principles of what is being studied." In discussing the *scientific method*, Shortley and Williams [2, p. 2] state: "The scientific method is the systematic attempt to construct theories that correlate wide groups of observed facts and are capable of predicting the results of future observations. Such theories are tested by controlled experimentation and are accepted only so long as they are consistent with all observed facts."

In many systems and processes of scientific and engineering interest, the geometry, boundary conditions, and physical phenomena are so complex that it is beyond our present technical capability to formulate satisfactory analytical or numerical models and approaches. In these cases, experimentation is necessary to define the behavior of the systems and/or processes (i.e., to find a solution to the problem).

1-1.2 Degree of Goodness and Uncertainty Analysis

If we are using property data or other experimentally determined information in an analytical solution, we should certainly consider how "good" the experimental information is. Similarly, anyone comparing results of a mathematical model with experimental data (and perhaps also with the results of other mathematical models) should certainly consider the *degree of goodness* of the data when drawing conclusions based on the comparisons. This situation is illustrated in Figure 1.2. In Figure 1.2*a* the results of two different mathematical models are compared with each other and with a set of experimental data. The authors of the two models might have a fine time arguing over which model compares better with the data. In Figure 1.2*b*, the same information is presented, but a range representing the uncertainty (likely amount of error) in the experimental value of Y has been plotted for each data point. It is immediately obvious that once the degree of goodness of the Y value is taken into consideration it is fruitless to argue for the validity of one model over another based only on how well the model results match the data. The "noise level" established by the data uncertainty effectively sets the resolution at which such comparisons can be made.

We will discuss such "validation" comparisons between simulation results and experimental results in considerable detail as we proceed. At this point, we will note that the experimental values of X will also contain errors, and so an uncertainty should also be associated with X. In addition, the simulation result also has uncertainty arising from modeling errors, errors in the inputs to the model, and possibly errors from the algorithms used to numerically solve the simulation equations.

From this example, one might conclude that even a person with no ambition to become an experimentalist needs an appreciation of the experimental process and the factors that influence the degree of goodness of experimental data and results from simulations.

Whenever the experimental approach is to be used to answer a question or to find the solution to a problem, the question of how good the results will be

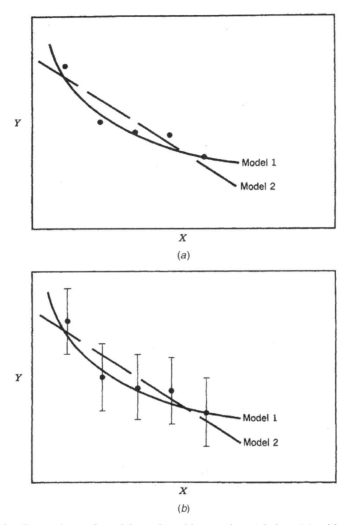

Figure 1.2 Comparison of model results with experimental data (*a*) without and (*b*) with consideration of uncertainty in *Y*.

should be considered long before an experimental apparatus is constructed and data are taken. If the answer or solution must be known within, say, 5% for it to be useful to us, it would make no sense to spend the time and money to perform the experiment only to find that the probable amount of error in the results was considerably more than 5%.

In this book we use the concept of *uncertainty* to describe the degree of goodness of a measurement, experimental result, or analytical (simulation) result. Schenck [3, p.7] quotes S. J. Kline as defining an experimental uncertainty as "what we think the error would be if we could and did measure it by calibration."

An error δ is a quantity that has a particular sign and magnitude, and a specific error δ_i is the difference caused by error source i between a quantity (measured or simulated) and its true value. As we will discuss in detail later, it is generally assumed that each error whose sign and magnitude are known has been removed by correction. Any remaining error is thus of unknown sign and magnitude,[1] and an uncertainty u is estimated with the idea that $\pm u$ characterizes the range containing δ.

Uncertainty u is thus an estimate: a $\pm u$ interval is an estimate of a range within which we believe the actual (but unknown) value of an error δ lies. This is illustrated in Figure 1.3, which shows an uncertainty interval $\pm u_d$ that contains the error δ_d whose actual sign and magnitude are unknown.

Uncertainty analysis (the analysis of the uncertainties in experimental measurements and in experimental and simulation results) is a powerful tool. This is particularly true when it is used in the planning and design of experiments. As we will see in Chapter 4, there are realistic, practical cases in which all the measurements in an experiment can be made with 1% uncertainty yet the uncertainty in the final experimental result will be greater than 50%. Uncertainty analysis, when used in the initial planning phase of an experiment, can identify such situations and save the experimentalist much time, money, and embarassment.

1-1.3 Experimentation and Validation of Simulations

Over the past several decades, advances in computing power, modeling approaches, and numerical solution algorithms have increased the ability of the scientific and engineering community to simulate real-world processes to the point that it is realistic for predictions from surprisingly detailed simulations to be used to replace much of the experimentation that was previously necessary to develop designs for new systems and bring them to the market. The new systems to which we refer cover the gamut from simple mechanical and structural devices to rocket engine injectors to commercial aircraft to military weapons systems to nuclear power systems.

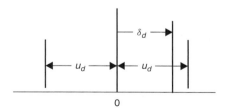

0

Figure 1.3 An uncertainty u defines an interval that is estimated to contain the actual value of an error of unknown sign and magnitude.

---

[1]There are asymmetric errors that are more likely to (or are certain to) have one sign rather than the other. Treatment of these by either "zero centering" or estimating asymmetric uncertainties is discussed in Chapter 5 and Appendix E.

In the past, it was necessary to test (experimentally determine) subsystem and system performance at numerous set points covering the expected domain of operation of the system. For large, complex systems the required testing program can be prohibitively expensive, if not outright impossible, with available finite resources. The current approach seeks to replace some or much of the experimentation with (cheaper) simulation results that have been validated with experimental results at selected set points—but to do this with confidence one must know "how good" the predictions are at the selected set points. This has led to the emergence of the field called verification and validation (V&V) of simulations (e.g., models, codes).

The verification part refers to application of approaches to determine that the algorithms solve the equations in the model correctly and to estimate the numerical uncertainty if the equations are discretized as, for example, in the finite-difference, finite-element, and finite-volume approaches used in computational mechanics. *Verification* addresses the question of whether the equations are solved correctly but does not address the question of how well the equations represent the real world. *Validation* is the process of determining the degree to which a model is an accurate representation of the real world—it addresses the question of how good the predictions are.

Verification is a necessary component of the validation process and will be described briefly with references cited to guide the reader who desires more detail, but more than that is beyond the scope of what we want to cover in this book. Since experimentation and the uncertainties in experimental results and in simulation results are central issues in validation, the details of validation *are* covered in this book. Basic ideas and concepts are developed as they arise naturally in the discussion of experimental uncertainty analysis—for example, estimating the uncertainty in the simulation result due to the uncertainties in the simulation inputs is discussed in Section 4-8. The application of the ideas and concepts in validation are covered in Chapter 6 with detailed discussion and examples.

1-2 EXPERIMENTAL APPROACH

1-2.1 Questions to Be Considered

When an experimental approach is to be used to find a solution to a problem, many questions must be considered. Among these are the following:

1. What question are we trying to answer? (What is the problem?)
2. How accurately do we need to know the answer? (How is the answer to be used?)
3. What physical principles are involved? (What physical laws govern the situation?)
4. What experiment or set of experiments might provide the answer?

5. What variables must be controlled? How well?
6. What quantities must be measured? How accurately?
7. What instrumentation is to be used?
8. How are the data to be acquired, conditioned, and stored?
9. How many data points must be taken? In what order?
10. Can the requirements be satisfied within the budget and time constraints?
11. What techniques of data analysis should be used?
12. What is the most effective and revealing way to present the data?
13. What unanticipated questions are raised by the data?
14. In what manner should the data and results be reported?

Although by no means all-inclusive, this list does indicate the range of factors that must be considered by the experimentalist. This might seem to be a discouraging and somewhat overwhelming list, but it need not be. With the aid of uncertainty analysis and a logical, thorough approach in each phase of an experimental program, the apparent complexities often can be reduced and the chances of achieving a successful conclusion enhanced.

A key point is to avoid becoming so immersed in the many details that must be considered that the overall objective of the experiment is forgotten. This statement may sound trite, but it is true nonetheless. We perform an experiment to find the answer to a question. We need to know the answer within some uncertainty, the magnitude of which is usually determined by the intended use of the answer. Uncertainty analysis is a tool that we use to make decisions in each phase of an experiment, always keeping in mind the desired result and uncertainty. Properly applied, this approach will guide us past the pitfalls that are usually not at all obvious and will enable us to obtain an answer with an acceptable uncertainty.

1-2.2 Phases of Experimental Program

There are numerous ways that a general experimental program can be divided into different components or phases. For our discussions in this book, we consider the experimental phases as planning, design, construction, debugging, execution, data analysis, and reporting of results. There are not sharp divisions between these phases—in fact, there is generally overlap and sometimes several phases will be ongoing simultaneously (as when something discovered during debugging leads to a design change and additional construction on the apparatus).

In the *planning phase* we consider and evaluate the various approaches that might be used to find an answer to the question being addressed. This is sometimes referred to as the *preliminary design phase*.

In the *design phase* we use the information found in the planning phase to specify the instrumentation needed and the details of the configuration of the experimental apparatus. The test plan is identified and decisions made on the ranges of conditions to be run, the data to be taken, the order in which the runs will be made, and so on.

During the *construction phase*, the individual components are assembled into the overall experimental apparatus, and necessary instrument calibrations are performed.

In the *debugging phase*, the initial runs using the apparatus are made and the unanticipated problems (which must always be expected!) are addressed. Often, results obtained in the debugging phase will lead to some redesign and changes in the construction and/or operation of the experimental apparatus. At the completion of the debugging phase, the experimentalist should be confident that the operation of the apparatus and the factors influencing the uncertainty in the results are well understood.

During the *execution phase*, the experimental runs are made and the data are acquired, recorded, and stored. Often, the operation of the apparatus is monitored using checks that were designed into the system to guard against unnoticed and unwanted changes in the apparatus or operating conditions.

During the *data analysis phase*, the data are analyzed to determine the answer to the original question or the solution to the problem being investigated.

In the *reporting phase*, the data and conclusions should be presented in a form that will maximize the usefulness of the experimental results.

In the chapters that follow we discuss a logical approach for each of these phases. We will find that the use of uncertainty analysis and related techniques (e.g., balance checks) will help to ensure a maximum return for the time, effort, and financial resources invested.

1-3 BASIC CONCEPTS AND DEFINITIONS

There is no such thing as a perfect measurement. All measurements of a variable contain inaccuracies. Because it is important to have an understanding of these inaccuracies if we are to perform experiments (use the experimental approach to answer a question) or if we are simply to use values that have been determined experimentally, we must carefully define the concepts involved.

1-3.1 Errors and Uncertainties

Consider a variable X in a process that is considered to be steady so that its true value (X_{true}) is constant. Measurements of the variable are influenced by a number of elemental error sources—such as the errors in the standard used for calibration and from an imperfect calibration process; errors caused by variations in ambient temperature, humidity, pressure, vibrations, electromagnetic influences; unsteadiness in the "steady-state" phenomenon being measured; errors due to undesired interactions of the transducer with the environment; errors due to imperfect installation of the transducer; and others. As an example, suppose that a measurement system is used to make N successive measurements of X and that the measurements are influenced by five significant error sources, as shown in Figure 1.4.

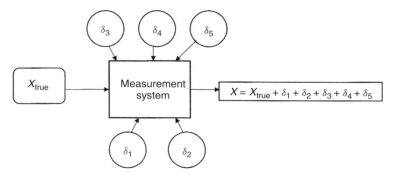

Figure 1.4 Measurement of a variable influenced by five error sources.

The first three of those measurements are given by

$$X_1 = X_{true} + (\delta_1)_1 + (\delta_2)_1 + (\delta_3)_1 + (\delta_4)_1 + (\delta_5)_1$$
$$X_2 = X_{true} + (\delta_1)_2 + (\delta_2)_2 + (\delta_3)_2 + (\delta_4)_2 + (\delta_5)_2 \qquad (1.1)$$
$$X_3 = X_{true} + (\delta_1)_3 + (\delta_2)_3 + (\delta_3)_3 + (\delta_4)_3 + (\delta_5)_3$$

where δ_1 is the value of the error from the first source, δ_2 the value of the error from the second source, and so on. Each of the measurements X_1, X_2, and X_3 has a different value since errors from some of the sources vary during the period when measurements are taken and so are different for each measurement while others do not vary and so are the same for each measurement. Using traditional nomenclature, we assign the symbol β (beta) to designate an error that does not vary during the measurement period and the symbol ϵ (epsilon) to designate an error that does vary during the measurement period. For this example, we will assume that the errors from sources 1 and 2 do not vary and the errors from sources 3, 4, and 5 do vary, so that Eqs. (1.1) can be written

$$X_1 = X_{true} + \beta_1 + \beta_2 + (\epsilon_3)_1 + (\epsilon_4)_1 + (\epsilon_5)_1$$
$$X_2 = X_{true} + \beta_1 + \beta_2 + (\epsilon_3)_2 + (\epsilon_4)_2 + (\epsilon_5)_2 \qquad (1.2)$$
$$X_3 = X_{true} + \beta_1 + \beta_2 + (\epsilon_3)_3 + (\epsilon_4)_3 + (\epsilon_5)_3$$

Since just by looking at the measured values we cannot distinguish between β_1 and β_2 or among ϵ_1, ϵ_2, and ϵ_3, Eq. (1.3) describes what we actually have,

$$X_1 = X_{true} + \beta + (\epsilon)_1 \quad X_2 = X_{true} + \beta + (\epsilon)_2 \quad X_3 = X_{true} + \beta + (\epsilon)_3 \quad (1.3)$$

where now

$$\beta = \beta_1 + \beta_2 \qquad \epsilon = \epsilon_3 + \epsilon_4 + \epsilon_5 \qquad (1.4)$$

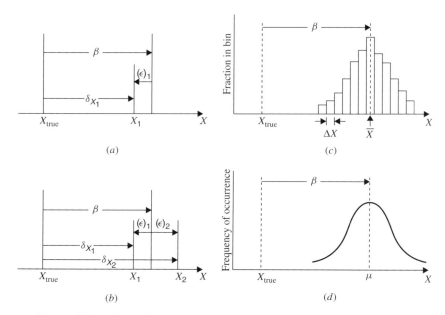

Figure 1.5 Effect of errors on successive measurements of a variable X.

This process of making successive measurements of X is shown schematically in Figure 1.5. In Figure 1.5a, the first measurement X_1 is shown. The difference between the measured value and the true value is the total error δ_{X_1}, which is the sum of the systematic error β (the combination of all of the errors from the systematic elemental error sources) and the random error $(\epsilon)_1$ (the combination at the time X_1 is measured of all of the errors from the error sources that vary during the period that our N measurements are taken). In Figure 1.5b the second measurement is also shown, and of course the total error δ_{X_2} differs from δ_{X_1} because the random error ϵ is different for each measurement.

If we continued to acquire additional measurements, we could plot a *histogram*, which presents the fraction of the N total measurements with values between X and $X + \Delta X$, $X + \Delta X$ and $X + 2\Delta X$, $X + 2\Delta X$ and $X + 3\Delta X$, and so on, versus X, where ΔX is the *bin width*. This is shown schematically in Figure 1.5c and allows us to view the *distribution* of the total of N measured values. This distribution of the *sample population* of N measurements tends to have a larger number of the measured values near the mean of the sample and a decreasing number of measured values as one moves away from the mean. As will be discussed in detail in Chapter 2, a *mean value* $\overline{X}$ can be calculated, as can a *standard deviation* s, which is an indicator of the width of the distribution of the X values (the amount of "scatter" of the measurements caused by the errors from the elemental sources that varied during the measurement period).

As the number of measurements approached infinity, the *parent population distribution* would likely appear as shown in Figure 1.5d, with the mean μ

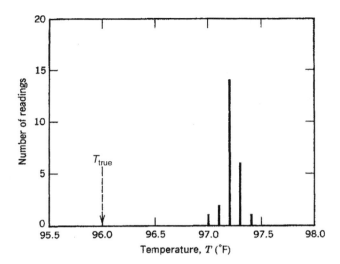

Figure 1.6 Histogram of temperatures read from a thermometer by 24 students.

offset from X_{true} by β, the combination of all of the systematic errors. Of course, we never have an infinite number of measurements, but conceptually the idea of the parent population distribution is very useful to us. The statistics of sample and parent distributions are discussed in Chapter 2.

An example of this behavior exhibited by a real set of measurements is shown in Figure 1.6. A thermometer immersed in an insulated container of water was read independently by 24 of our students to the nearest tenth of a degree Fahrenheit. Unknown to the students, the thermometer was biased ("read high") by a little over a degree, and the "true" temperature of the water was about 96.0°F. The temperatures read by the students are distributed around an average value of about 97.2°F and are biased (offset) from the true value of 96.0°F.

With such a data sample, what we would like to do is use information from the sample to specify some range ($X_{\text{best}} \pm u_X$) within which we think X_{true} falls. Generally we take X_{best} to be equal to the average value of the N measurements (or to the single measured value X if $N = 1$). The uncertainty u_X is an estimate of the interval ($\pm u_X$) that likely contains the magnitude of the combination of all of the errors affecting the measured value X. Look back at the first measurement of X illustrated in Figure 1.5a and imagine that it is influenced by five significant error sources as in Figure 1.4. Then, recalling Eq. (1.2), the expression for the first measured value X_1 is given by

$$X_1 = X_{\text{true}} + \beta_1 + \beta_2 + (\epsilon_3)_1 + (\epsilon_4)_1 + (\epsilon_5)_1 \qquad (1.5)$$

To associate an uncertainty with a measured X value, we need to have elemental uncertainty estimates for all of the elemental error sources. That is, u_1 is an uncertainty that defines an interval ($\pm u_1$) within which we think the value of

β_1 falls, while u_3 is an uncertainty that defines an interval ($\pm u_3$) within which we think the value of ϵ_3 falls.

Using the concepts and procedures in the ISO *Guide to the Expression of Uncertainty in Measurement* [4], a *standard uncertainty* (u) is defined as an estimate of the standard deviation of the parent population from which a particular elemental error originates. Then u_X is found from the combination of all of the elemental standard uncertainties as

$$u_X = (u_1^2 + u_2^2 + u_3^2 + u_4^2 + u_5^2)^{1/2} \tag{1.6}$$

For N measurements of X, the standard deviation s_X of the sample distribution shown in Figure 1.5c can be calculated as

$$s_X = \left[\frac{1}{N-1} \sum_{i=1}^{N} (X_i - \overline{X})^2 \right]^{1/2} \tag{1.7}$$

where the mean value of X is calculated from

$$\overline{X} = \frac{1}{N} \sum_{i=1}^{N} X_i \tag{1.8}$$

How can we determine which significant error sources' influences are included in s_X and which ones are not? There is only one apparent answer—we must determine which of the elemental error sources did not vary during the measurement period and thus produced errors that were the same in each measurement. These are the *systematic* error sources, and their influence is not included in s_X. The influences of all of the elemental error sources that varied during the measurement period (whether one knows the number of them or not) are included in s_X.

To understand and to take into account the effects of all of the significant error sources, then, we must identify two categories—the *systematic* sources whose effects are not included in s_X and the so-called *random* sources that varied during the measurement period and whose effects are included in s_X. (Assignment of an error source to the random category does not imply that the variation in the errors from that source is truly random. As will be discussed later, the *combination* of the errors from the different "random" sources tends to have a random behavior.)

In the case being considered, s_X reflects the contributions from the variable errors from sources 3, 4, and 5 and is related to the standard uncertainties from those elemental sources through

$$s_X = (u_3^2 + u_4^2 + u_5^2)^{1/2} \tag{1.9}$$

and so Eq. (1.6) becomes

$$u_X = (u_1^2 + u_2^2 + s_X^2)^{1/2} \tag{1.10}$$

This leaves the standard uncertainties for the systematic error sources to be estimated before we can determine the standard uncertainty u_X to associate with the measured variable X. In this book, as in the American National Standards Institute/American Society of Mechanical Engineers (ANSI/ASME) standard PTC19.1-2005, *Test Uncertainty* [5], the systematic standard uncertainties are designated with the symbol b_i, which is understood to be an estimate of the standard deviation of the distribution of the parent population from which a particular systematic error β_i originates. (This b nomenclature has its origin in the decades-past use of the word *bias*, which has been replaced in current usage with the word *systematic*.)

Upon first encountering the idea of a "parent population" for a systematic error, the idea itself might seem to be self-contradictory. A simple example serves to put the meaning into perspective, however. Imagine that we plan to use a Coleman–Steele model 22 voltmeter to record a measurement system output of 4.360 V and postulate that there are no sources of random error or unsteadiness. We enter a room in which there are 1000 model 22 voltmeters, and we choose the 14th one. In a perfect calibration apparatus we input 4.360 V into the voltmeter and its output is 4.363 V—it "reads high" by +0.003 V (which is the value of its systematic error, β_{14}). Next we choose the 29th model 22 voltmeter, repeat the calibration process, and find that it "reads low" by −0.010 V (which is the value of its systematic error, β_{29}). We repeat this for the 45th, 73rd, 102nd, . . . voltmeters and end up with a *distribution* of β's from which we can calculate a standard deviation b. This value b would be the systematic uncertainty that we would associate with the voltage measurement of approximately 4.360 V from the model 22 voltmeter used in our measurement system, *because we do not know what the actual β is for that particular meter that we happened to choose from the available population of model 22 voltmeters.*

The systematic standard uncertainties for the elemental sources are estimated in a variety of ways that are discussed and illustrated in later sections. Among the ways used to obtain estimates are use of previous experience, manufacturer's specifications, calibration data, results from specially designed "side" experiments, results from analytical models, and others. Using this nomenclature, Eq. (1.10) is then written as

$$u_X = (b_1^2 + b_2^2 + s_X^2)^{1/2} \qquad (1.11)$$

The ISO guide [4, p. 2] recommends designation of the standard uncertainties for the elemental sources by the way in which they are estimated. A *type A* evaluation of uncertainty is defined as a "method of evaluation of uncertainty by the statistical analysis of series of observations," and the symbol s is used. A *type B* evaluation of uncertainty is defined as a "method of evaluation of uncertainty by means other than the statistical analysis of series of observations," and the generic symbol u is used. The recommendations in the guide [4] and their use and implementation in this text and in Ref. 5 are discussed in detail later.

If in the case under discussion b_1 was estimated by a statistical evaluation using calibration data, it would be a *type A standard uncertainty* and would be designated $b_{1,A}$. If b_2 was estimated using an analytical model of the transducer and its boundary conditions, it would be a *type B standard uncertainty* and would be designated $b_{2,B}$. If s_X was calculated statistically as described, it would be a *type A standard uncertainty* and Eq. (1.11) would become

$$u_X = (b_{1,A}^2 + b_{2,B}^2 + s_{X,A}^2)^{1/2} \qquad (1.12)$$

In summary, an *error* is a quantity with a given sign and magnitude. We presume that corrections will be or have been made for errors of *known* sign and magnitude (more discussion on this later). So in the remainder of this book, unless specifically stated, when we refer to an error, it is of *unknown* sign and magnitude and presumed equally probable to be positive or negative. An *uncertainty* u is an estimate of an interval ($\pm u$) that likely contains the magnitude of the error. Since the standard uncertainty defined in the ISO guide [4] and used in Ref. 5 and this book requires no assumption about the form of the parent population error distribution, the probability that the magnitude of the error falls within $\pm u$ is not known. This is discussed in the following section.

1-3.2 Degree of Confidence and Uncertainty Intervals

We stated earlier that we would like to take the information from a data sample to specify some range ($X_{\text{best}} \pm u_X$) within which we think X_{true} lies. The problem, as just noted, with using the standard uncertainty u_X for this range is that no probability can be associated with it. What is needed is an expanded uncertainty estimate, U_X, so that we can say that we are C percent confident that the true value of X lies within the interval

$$X_{\text{best}} \pm U_X$$

We usually assume X_{best} to be the mean value of the N measurements we have made (or if $N = 1$, *the* value of the single measurement), and U_X is the uncertainty in X that corresponds to our estimate with C percent confidence of the effects of the combination of the systematic and random errors. That is, $\pm U_X$ is the range within which we estimate δ (the value of the total error) lies C percent of the time. If, for example, we make a 95% confidence estimate of U_X, we would expect that X_{true} would be in the interval $X_{\text{best}} \pm U_X$ about 95 times out of 100. The methodology for obtaining the expanded uncertainty U_X from the standard uncertainty u_X is described in Chapter 2.

The confidence specification is necessary because we have made an estimate. We can always be 100% confident that the true value of some quantity will lie between plus and minus infinity, but specifying U_X as infinite provides no useful information to anyone. It is not necessary to perform an experiment to find that result!

The idea of degree of confidence in an uncertainty specification is vividly and humorously illustrated in an anecdote reported by Abernethy et al. [6, p. 162]:

> In the 1930's, P. H. Myers at NBS and his colleagues were studying the specific heat of ammonia. After several years of hard work, they finally arrived at a value and reported the result in a paper. Toward the end of the paper, Myers declared: "We think our reported value is good to one part in 10,000; we are willing to bet our own money at even odds that it is correct to two parts in 10,000; furthermore, if by any chance our value is shown to be in error by more than one part in 1000, we are prepared to eat our apparatus and drink the ammonia!"

1-3.3 Expansion of Concept from "Measurement Uncertainty" to "Experimental Uncertainty"

Sometimes the uncertainty specification must correspond to more than an estimate of the "goodness" with which we can measure something. This is true for cases in which the quantity of interest has a variability unrelated to the errors inherent in the measurement system. When discussing this, it is helpful to consider three broad categories of experiments: timewise experiments at a (supposed) steady-state condition, sample-to-sample experiments, and transient experiments (in which there is a time-varying process).

In a typical *timewise experiment* we might want to measure a fluid flow rate in a system operating at a steady-state condition. Although the system might be in "steady" operation, there inevitably will be some time variations of flow rate that will appear as random errors in a series of flow rate measurements taken over a period of time. In addition, the inability to reset the system at exactly the same operating condition from trial to trial will cause additional data scatter.

In *sample-to-sample experiments*, measurements are made on multiple samples so that in a sense sample identity corresponds to the passage of time in timewise experiments. In sample-to-sample experiments, the variability inherent in the samples themselves causes variations in measured values in addition to the random errors in the measurement system.

As a practical example of such a case, suppose that we have been retained to determine the heating value of lignite in a particular geographic region. Lignite is a soft, brown coal with relatively low heating value and high ash and moisture content. However, a utility company is interested in the possibility of locating a new lignite-fueled power plant in the midst of a region (shown in Figure 1.7) that contains several large deposits of lignite. The savings in coal transportation costs might offset the negative effects of the low heating value of lignite and make such a power plant economically feasible. A decision on the economic feasibility depends critically, of course, on the lignite heating value used in the calculations.

The heating value of a 1-g sample of lignite can be determined with an uncertainty of less than 2% in the laboratory using a bomb calorimeter. However, because of large variations in the composition of the lignite within a single deposit and from deposit to deposit, the heating values determined for samples

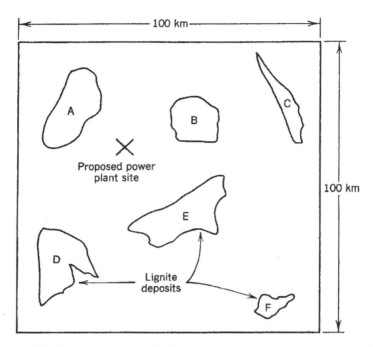

Figure 1.7   Region containing lignite deposits and potential power plant site.

from the northern and southern portions of deposit C, for instance, might differ by 20% or more. Even larger differences might very well be observed when results are compared for samples from entirely different deposits.

In this situation, the variation within a set of individual measurements of lignite heating value is not due primarily to errors in the measurements. It is due to the natural variation of the physical variable that is being measured. The amount of the variation within a set of measurements will differ depending on whether the set is taken from within a small area of a single deposit, from the entire area of a single deposit, or from different deposits. We might conclude from this example that the answer to the question, "What is the uncertainty in the lignite heating value?" depends on what is meant when one says, "the lignite." The answer will be one thing for a single 1-g sample and quite another thing for deposits A, B, and E considered together. Thus we can conclude that the uncertainty estimate depends on what the question is.

In *transient experiments*, the process of interest varies with time and there are some unique circumstances involved in obtaining multiple measurements of a variable at a given condition. As an example of one type of transient test, consider measuring the thrust of a rocket engine with a single load cell from the instant of ignition ($t = 0$) until the time when the desired "steady" operating condition is reached. Only one measurement of thrust can be made at $t = 0.3078$ s in a given test. To obtain a sample population of measurements of thrust at

$t = 0.3078$ s, the engine must be tested again, and again, and again. This type of test has obvious similarities to the sample-to-sample category, and even more so if multiple engines of the same design are considered as part of the sample population of measurements.

A second type of transient test is when the process is periodic. Consider measuring the temperature at a point within a cylinder of an internal combustion engine operating at a steady condition. This type of test has some obvious similarities to the timewise steady category discussed earlier. Multiple measurements of temperature at the "same" operating point can be obtained by making a measurement at 30° before top dead center, for example, over a number of repetitions of the periodic process.

Measurements in a transient test typically have elemental error sources in addition to those normally present in a timewise steady test. Imperfect dynamic response of the measurement system can result in errors in the magnitude and phase of the measured quantity. An introduction to this subject is presented in Appendix F.

1-3.4 Elemental Systematic Errors and Effects of Calibration

Over the years, the misconception has persisted that the effects of systematic errors can be removed by calibration. In fact, systematic errors have often been summarily dismissed in books and articles on error analysis or uncertainty analysis by a simple statement that "all bias errors have been eliminated by calibration." In real testing situations, this is never the case.

Consider as an example the measurement of a temperature using a thermocouple (tc) system as shown in Figure 1.8a. The tc is connected to a data acquisition system (das) consisting of an electronic reference junction, signal conditioning, an analog-to-digital converter, and a digital voltmeter. The tc is exposed to some temperature T_{true} that one wishes to "measure" and the system output is a voltage E.

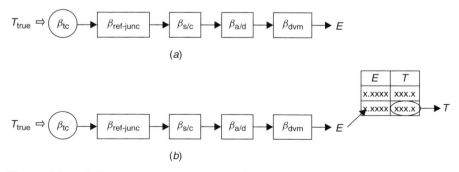

Figure 1.8 (a) Thermocouple system with its output voltage E. (b) No individual calibration—used with generic temperature–voltage table to "measure" a temperature.

Suppose that the thermocouple is used as supplied and is not individually calibrated. As shown in Figure 1.8b, the voltage E is used in a generic T-versus-E table for the particular type of thermocouple and the corresponding temperature T is found—this is the temperature that is said to be the measured value of T_{true}. The uncertainty in this value includes contributions from the elemental systematic errors:

- β_{tc}, the amount this tc differs from the generic tc in the table
- $\beta_{\text{ref-junc}}$, the error from the electronic reference junction
- $\beta_{\text{s/c}}$, the error from the signal conditioner
- $\beta_{\text{a/d}}$, the error from the analog to digital converter
- β_{dvm}, the error from the digital voltmeter

Now suppose the thermocouple is calibrated as shown in Figure 1.9a. The thermocouple and the temperature standard are both exposed to the same temperature T_{true}, and the voltage E output by the thermocouple system (tc + das)

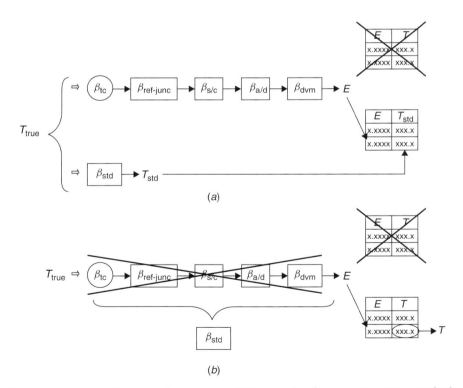

(a)

(b)

Figure 1.9 (a) Calibration of tc system of Figure 1.8 using a temperature standard. (b) Use of calibrated system to measure a temperature.

and the temperature T_{std} indicated by the standard are entered as a data pair in the calibration table, which now replaces the generic table used previously.

When the tc system is now used to measure a value T_{true}, as shown in Figure 1.9b, the voltage output E is used to enter the calibration table and the corresponding temperature retrieved is effectively that which would be indicated by the standard. The uncertainty in the resulting temperature T now includes the contribution from the systematic error β_{std} (the amount T_{std} differs from T_{true}), which *replaces* the contributions from the errors β_{tc}, $\beta_{ref\text{-}junc}$, $\beta_{s/c}$, $\beta_{a/d}$, and β_{dvm}.

So we see that some systematic errors can be replaced by (hopefully!) smaller ones by application of the calibration process. In this example, we have not considered other systematic error sources that might be present in the test (such as installation effects, interaction of the tc with the environment, and others that we will discuss in detail in later chapters) that are not present in the well-controlled calibration process.

It is also critical to realize just what is being calibrated. If the tc and data acquisition system that are calibrated together are used in the test, then the error situation is as described above. However, if the data acquisition system used in the calibration (das_{cal}) is replaced by another (das_{test}) for the test, the systematic error from the standard β_{std} then *only* replaces the error contribution β_{tc}. The uncertainty in the resulting temperature T now includes contributions from (1) β_{std}; (2) $\beta_{ref\text{-}junc}$, $\beta_{s/c}$, $\beta_{a/d}$, and β_{dvm} from das_{cal}; *and* (3) $\beta_{ref\text{-}junc}$, $\beta_{s/c}$, $\beta_{a/d}$, and β_{dvm} from das_{test}.

1-3.5 Repetition and Replication

In discussing errors and uncertainties in measurements, it is useful to draw a careful distinction between the meanings of the words *repetition* and *replication*. We will use repetition in its common sense, that is, to mean that something is repeated. When we use the word replication, we will be specifying that the repetition is carried out in a very specific manner. The reason for doing this is that different factors influence the errors in a series of measurements, depending on how the repetition is done. The idea of considering uncertainties at different orders of replication level was suggested by Moffat [7–9] for the timewise category of experiments, and we will also find it useful for sample-to-sample experiments. As did Moffat, we will find it convenient to define three different levels of replication: zeroth order, first order, and Nth order.

At the *zeroth-order replication level* in a timewise experiment, we hypothesize that time is "frozen" so that the process being measured is absolutely steady. This allows only the variations inherent in the measuring system itself to contribute to random errors. In a sample-to-sample type of experiment, this corresponds to consideration of a single fixed sample.

If we make a zeroth-order estimate of the systematic and random uncertainties associated with an instrument, this is the "best" we would ever be able to achieve if we used that instrument. If this zeroth-order estimate shows that the uncertainties are larger than those that are allowable in our experiment, it is obvious that we should not perform the experiment using that instrument.

In considering uncertainties at the *first-order replication level* in a timewise experiment, we hypothesize that time runs but all instrument identities are fixed. When estimating uncertainties at this level of replication, the variability of the experimental measurement is influenced by all of the factors that contribute to unsteadiness during repeated trials with the experimental apparatus. Depending on the length of time covered by a replication at the first-order level, different factors such as variations of humidity, barometric pressure, ambient temperature, and the like can influence the random portion of the experimental uncertainty. In addition, our inability to reproduce a set point exactly with the experimental apparatus influences the variations in experimental results.

A first-order estimate of the random uncertainty is indicative of the *scatter* that we would expect to observe during the course of a timewise experiment. If on repeated trials over the course of time we observe a scatter significantly larger than that given by the first-order estimate, it is very likely that there is some effect that has not been taken into account and that we should probably investigate further. This use of first-order random uncertainty estimates in debugging timewise experiments is discussed in detail in Chapter 5.

In considering uncertainties at the first-order replication level in a sample-to-sample experiment, we imagine that all instruments remain the same as sample after sample is tested. Additional variations observed above the level of random errors at the zeroth order are indicative of the variations in the samples themselves.

When considering what uncertainties should be specified when we speak of where the true value lies relative to our measurements, we make estimates at the *Nth-order replication level*. Such an Nth-order estimate includes the first-order replication level estimate of random uncertainty combined with a systematic uncertainty estimate that considers all of the systematic error sources that influence our measurements. Using Moffat's concept for timewise steady experiments, at the Nth-order replication level both time and instrument identities are considered to vary. At this level, we hypothesize that after each measurement each instrument is considered to have been replaced by another of the same type, model number, and so on.

What this essentially means is that the systematic error associated with a particular instrument becomes a random variable when the instrument identity is allowed to vary. To see this, consider that a particular model of pressure gauge might be specified by the manufacturer as "accurate within $\pm 1\%$ of full scale." The particular gauge we use might actually read high by 0.75% of full scale (a systematic error). If we replace that gauge by an "identical" gauge, it might very well read low by 0.37% of full scale. Thus, when we consider replacing each instrument with an identical instrument for each reading, the systematic error associated with the instrument becomes a random error at the Nth-order level of replication. Using an Nth-order model for the systematic error, then, we can take the view that a particular systematic error is a single realization drawn from some parent population of possible systematic errors. This concept is often useful when formulating a systematic uncertainty estimate.

From the preceding discussion it should be evident that the reason for considering different levels of replication is that the factors that are considered to influence the uncertainty in the experiment differ with the level of replication. We shall make use of this point as we consider the analysis and design of experiments in later chapters.

1-4 EXPERIMENTAL RESULTS DETERMINED FROM MULTIPLE MEASURED VARIABLES

In many, if not most, experiments or testing programs the experimental result is not directly measured but is determined by combining multiple measured variables in a *data reduction equation* (DRE). Examples are the dimensionless groups, such as drag coefficient, Nusselt number, Reynolds number, and Mach number, that are often used to present the results of a test. Other examples are the density of a gas (the result) determined using the ideal gas expression

$$\rho = \frac{P}{RT} \tag{1.13}$$

and measurements of pressure P and temperature T, or a heat flux (result) on the inside of a wall using an inverse heat conduction method and measurements of temperature versus time on the outside of the wall.

When a DRE is used, we must consider how the systematic and random uncertainties in the measured variables *propagate* through the DRE to produce the systematic and random uncertainties associated with the result. This is discussed in Chapter 3. There are two propagation approaches in common use today—the Taylor Series Method (TSM) and the Monte Carlo Method (MCM). Both are described and their applications illustrated in the following chapters.

In some experiments, the objective is to determine the relationship among several results. An example is determining the resistance characteristics of a rough-walled pipe in a flow system, as shown schematically in Figure 1.10.

A classic experiment of this kind was executed and reported by Nikuradse [10], who investigated the effect of wall roughness on fully developed flow in circular pipes that had been roughened by gluing a specific size of sand onto the interior wall of a given pipe test section. The results of that experimental investigation were reported with friction factor λ as a function of Reynolds number Re and "relative sand-grain roughness" r/k, as shown in Figure 1.11.

Using modern nomenclature friction factor is defined as

$$f = \frac{\pi^2 d^5 (\Delta P)}{8 \rho Q^2 L} \tag{1.14}$$

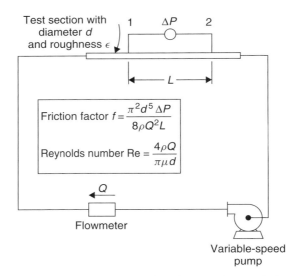

Figure 1.10 Experimental determination of resistance characteristics of rough-walled pipe.

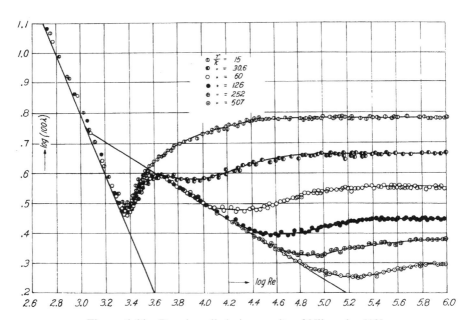

Figure 1.11 Rough-walled pipe results of Nikuradse [10].

and the Reynolds number as

$$Re = \frac{4\rho Q}{\pi \mu d} \tag{1.15}$$

and the third nondimensional group is ϵ/d, where ϵ in this example is taken as the "equivalent sand-grain roughness" (which is a single length scale descriptor of the roughness). In such an experiment, values for the following variables are found by measurement or from reference sources (property tables, dimensioned drawings):

- d, the pipe diameter
- ϵ, the roughness "size"
- L, the distance between pressure taps
- ΔP, the directly measured pressure drop over distance L
- Q, the volumetric flow rate of the fluid
- ρ, the density of the fluid
- μ, the dynamic viscosity of the fluid

and the experimental results f, Re, and ϵ/d are determined from DREs.

If one were going to test a pipe with a new kind of wall roughness and results such as those shown in Figure 1.11 were not yet known, in the planning phase of the experiment one would use *general uncertainty analysis* (discussed in Chapter 4) to investigate the overall uncertainty behavior of the DREs—Eqs. (1.14) and (1.15) and ϵ/d. Once satisfied that the results (f, Re, ϵ/d) could be obtained with an acceptable uncertainty, then in the postplanning phases described in Section 1-2.2 one would investigate the detailed behavior of the systematic and random uncertainties using *detailed uncertainty analysis* (discussed in Chapter 5).

At the completion of the experimental program, one might want to show the relationship among the results (in addition to a plot similar to Figure 1.11) by using a regression equation such as that [11] which represents the famous Moody diagram:

$$\frac{1}{f^{0.5}} = -1.8 \log_{10} \left[\frac{6.9}{Re_D} + \left(\frac{\epsilon/D}{3.7} \right)^{1.11} \right] \tag{1.16}$$

When such an expression is used to obtain a value of the result—in this case friction factor f—the uncertainty which should be associated with that value should take into account the uncertainties in the data points used to obtain the regression. The approach to do this is discussed in detail in Chapter 7.

1-5 GUIDES AND STANDARDS

1-5.1 Experimental Uncertainty Analysis

In 1993, the *Guide to the Expression of Uncertainty in Measurement* was published by the International Organization for Standardization (ISO) in its name and those of six other international organizations.[2] According to the foreword in the ISO guide, "In 1977, recognizing the lack of international consensus on the expression of uncertainty in measurement, the world's highest authority in metrology, the Comite International des Poids et Mesures (CIPM), requested the Bureau International des Poids et Mesures (BIPM) to address the problem in conjunction with the national standards laboratories and to make a recommendation." After several years of effort, this led to the assignment of responsibility to the ISO Technical Advisory Group on Metrology, TAG 4, which then established Working Group 3 to develop a guidance document. This ultimately culminated in the publication of the ISO guide, which has been accepted as the de facto international standard for the expression of uncertainty in measurement (and is commonly referred to as the GUM).

As discussed previously in this chapter, the GUM recommends categorizing uncertainties as type A or B depending on the method of evaluation [statistical (A) or otherwise (B)]. The GUM discourages the use of systematic and random but then also states that sometimes dividing uncertainties into those that contribute to variability and those that do not can be useful. In fact, as discussed previously (Section 1-3.1), we cannot envision any way to identify all of the elemental error sources other than to divide them into categories that (1) vary and contribute to the standard deviation s (random standard uncertainty) and those that (2) do not vary and thus must be estimated with a systematic standard uncertainty b. In this book, we use this nomenclature, as does the ASME standard PTC 19.1-2005 [5].

The approach in the GUM is based on a Taylor Series Method (TSM) to model the propagation of uncertainties, and we will use TSM to denote this approach in this book. This is to distinguish that propagation approach from the Monte Carlo Method (MCM) of propagation presented in the recently published [12] Joint Committee for Guides in Metrology (JCGM) supplement.

From the foreword to that document:

In 1997 a Joint Committee for Guides in Metrology (JCGM), chaired by the Director of the Bureau International des Poids et Mesures (BIPM), was created by the seven international organizations that had originally in 1993 prepared the "Guide to the expression of uncertainty in measurement" (GUM) and the "International vocabulary of basic and general terms in metrology" (VIM). The JCGM assumed responsibility for those two documents from the ISO Technical Advisory Group 4 (TAG4).... JCGM has two Working Groups. Working Group 1, "Expression of

[2]Bureau International des Poids et Mesures (BIPM), International Electrotechnical Commission (IEC), International Federation of Clinical Chemistry (IFCC), International Union of Pure and Applied Chemistry (IUPAC), International Union of Pure and Applied Physics (IUPAP), and International Organization of Legal Metrology (OIML).

uncertainty in measurement", has the task to promote the use of the GUM and to prepare Supplements and other documents for its broad application . . . . Supplements such as this one are intended to give added value to the GUM by providing guidance on aspects of uncertainty evaluation that are not explicitly treated in the GUM. The guidance will, however, be as consistent as possible with the general probabilistic basis of the GUM.

The MCM requires estimation of probability distributions rather than simply standard uncertainties (standard deviations) as in the TSM. This, of course, is thoroughly discussed as the method is described and implemented in this book.

1-5.2 Validation of Simulations

The approach to validation of simulations described in this book is that in the recently issued [13] ASME standard V&V20-2009, *Standard for Verification and Validation in Computational Fluid Dynamics and Heat Transfer*. The approach is based on the concepts from experimental uncertainty analysis as presented in this book, the GUM, and JCGM 101:2008 and uses both the TSM and MCM techniques. From the foreword:

> The objective of this Standard is the specification of a verification and validation approach that quantifies the degree of accuracy inferred from the comparison of solution and data for a specified variable at a specified validation point. The scope of this Standard is the quantification of the degree of accuracy for cases in which the conditions of the actual experiment are simulated. Consideration of the accuracy of simulation results at points within a domain other than the validation points, for example interpolation/extrapolation in a domain of validation, is a matter of engineering judgment specific to each family of problems and is beyond the scope of this Standard. ASME PTC 19.1-2005 "Test Uncertainty" is considered a companion document to this Standard.

Previously both the American Institute of Aeronautics and Astronautics (AIAA) [14] and the ASME [15] published V&V guides that present the philosophy and procedures for establishing a comprehensive validation program, but neither provides approaches to quantitative evaluations of the comparison of the validation variables predicted by simulation and determined by experiment.

V&V is an area of current research, and there is (and likely will never be) no universally accepted single method for use in simulations in all disciplines ranging from computational mechanics to war gaming. Organizations other than the AIAA and ASME [e.g., 16] are currently in the process of developing V&V guides and standards.

1-6 A NOTE ON NOMENCLATURE

It is worthwhile at this point to emphasize several points on the nomenclature used in experimental uncertainty analysis and the way it has developed over the past several decades. In Appendix B, the authors outline the historical development of uncertainty analysis from the 1950s through the 1990s, and the reader is referred to it as a fairly concise overview.

For those familiar with the nomenclature used in books and the literature in the recent past (particularly in the United States), perhaps the most prominent change is that the symbols B and P are no longer used. Originally B was termed the "bias limit" and later called the "systematic uncertainty," and P was termed the "precision limit" and later called the "random uncertainty." These are no longer needed, as the concepts of u, the standard uncertainty; b, the systematic standard uncertainty; and s, the random standard uncertainty are unambiguous in their definitions and in their use in both the TSM and MCM for uncertainty propagation.

REFERENCES

1. *Webster's New Twentieth Century Dictionary*, 2nd ed., Simon & Schuster, New York, 1979.
2. Shortley, G., and Williams, D., *Elements of Physics*, 4th ed., Prentice-Hall, Upper Saddle River, NJ, 1965.
3. Schenck, H., *Theories of Engineering Experimentation*, 3rd ed., McGraw-Hill, New York, 1979.
4. International Organization for Standardization (ISO), *Guide to the Expression of Uncertainty in Measurement*, ISO, Geneva, 1993. Corrected and reprinted, 1995.
5. American National Standards Institute/American Society of Mechanical Engineers (ASME), *Test Uncertainty*, PTC-19.1-2005, ASME, New York, 2006.
6. Abernethy, R. B., Benedict, R. P., and Dowdell, R. B., "ASME Measurement Uncertainty," *Journal of Fluids Engineering*, Vol. 107, June 1985, pp. 161–164.
7. Moffat, R. J., "Contributions to the Theory of Single-Sample Uncertainty Analysis," *Journal of Fluids Engineering*, Vol. 104, June 1982, pp. 250–260.
8. Moffat, R. J., "Using Uncertainty Analysis in the Planning of an Experiment," *Journal of Fluids Engineering*, Vol. 107, June 1985, pp. 173–178.
9. Moffat, R. J., "Describing the Uncertainties in Experimental Results," *Experimental Thermal and Fluid Science*, Vol. 1, Jan. 1988, pp. 3–17.
10. Nikuradse, J. "Stromugsgestze in Rauhen Rohren," VDI Forschungsheft, No. 361 1950 (English Translation, NACA TM 1292).
11. White, F. M., *Fluid Mechanics*, 6th ed., McGraw-Hill, New York, 2008.
12. Joint Committee for Guides in Metrology (JCGM), "Evaluation of Measurement Data—Supplement 1 to the 'Guide to the Expression of Uncertainty in Measurement'—Propagation of Distributions Using a Monte Carlo Method," JCGM 101: 2008, France, 2008.
13. American Society of Mechanical Engineers (ASME), *Standard for Verification and Validation in Computational Fluid Dynamics and Heat Transfer*, V&V20-2009, ASME, New York, 2009.
14. American Institute of Aeronautics and Astronautics (AIAA), *Guide for the Verification and Validation of Computational Fluid Dynamics Simulations*, AIAA G-077-1998, AIAA, New York.

15. American Society of Mechanical Engineers (ASME), *Guide for Verification and Validation in Computational Solid Mechanics*, ASME V&V10-2006, ASME, New York, 2006.

16. NASA, "Standard for Models and Simulations," NASA-STD-7009, Draft, 2008.

PROBLEMS

1.1 Consider the lignite heating value example discussed in Section 1-3.3. A salesman drops by your office and says that he can supply a calorimetry system that can determine the heating value of a 1-g sample of lignite with an uncertainty of about 1% as opposed to the 2% obtained using the present system. Is this idea worth considering if you need to find the heating value of the lignite in deposit E?

1.2 A Pitot-static probe is mounted to monitor the exit velocity of air exhausting from a duct in a process unit. The differential pressure output from the probe is applied to an analog pressure gauge, and a technician reads the gauge and notes the reading every hour. The operating condition of the process unit is held as steady as possible by its control system. If the gauge is located in a room with a conditioned environment, list the possible factors involved in causing scatter in the readings. What additional factors are involved if the gauge is located in a room where the temperature and humidity are uncontrolled? If the process unit is shut down and then restarted and reset to the same operating condition? If the gauge is replaced by another of the same model number from the same manufacturer?

2

ERRORS AND UNCERTAINTIES IN A MEASURED VARIABLE

In Chapter 1, the concepts of random and systematic errors were discussed and a quantity called the standard uncertainty u [1] was defined as an estimate of the standard deviation of the parent population from which a particular error originates. We also considered the combination of the effects of random and systematic errors and the concept of an uncertainty interval at a given level of confidence. All of these concepts are rooted in the study of the statistics of errors. Statistics is the study of mathematical properties of distributions of numbers (such as multiple temperature measurements in the thermometer example discussed in Chapter 1).

In this chapter we first discuss basic statistical concepts. Then we apply them to estimate the effects of random errors on the uncertainty of a variable. These concepts are extended to the estimation of systematic error effects and the determination of the total uncertainty for a measured variable.

2-1 STATISTICAL DISTRIBUTIONS

Consider the output of a pressure transducer that is monitored over a period of time while the transducer is supposedly measuring a "constant" input pressure. The measurements that are taken might appear as shown in the histogram plotted in Figure 2.1. The measurements scatter about a central value of about 86 mV, with some being higher and some lower. If additional measurements were taken so that the total number approached infinity, the resulting histogram curve would become smoother and more bell shaped, as shown in Figure 2.2.

The distribution of readings defined if an infinite number of samples could be taken is called the *parent distribution* or *parent population*. In reality, of

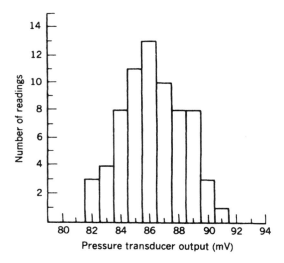

Figure 2.1   Histogram of measurement of output of a pressure transducer.

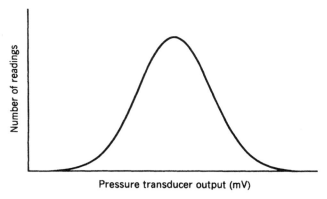

Figure 2.2   Distribution of measurements of output of a pressure transducer as the number of measurements approaches infinity.

course, we never have the time or resources to obtain an infinite number of measurements, so we must work with a *sample distribution* composed of a finite number of measurements taken from the parent population.

In the example shown in Figures 2.1 and 2.2, the scatter of the measurements is representative of the effects of the random errors. For purely random errors, the resulting infinite distribution will approach a Gaussian parent distribution such as that shown in Figure 2.2. Of course the pressure transducer measurements will also have systematic errors which affect all the measurements equally and are not detectable by taking multiple measurements.

As we discussed in Chapter 1, the standard uncertainties associated with the effects of the random and systematic errors are the estimates of the standard deviations of the parent populations for the error sources. For the random errors, this quantity is s_X, or the estimate of the standard deviation of the parent population for the measurements of X. For random errors we usually assume that this distribution is Gaussian, or normal.

For systematic errors that have not been corrected by calibration, we do not know the parent distributions for the possible values for these errors. The systematic error from a specific source will be fixed for a given measurement, but its sign and magnitude are unknown. We assume that this error is a single realization from some distribution of possible values. What we need to quantify the effects of the systematic error is the standard deviation of this distribution. There can be situations where the distribution might be approximated as Gaussian and other times when a rectangular, triangular, or other distribution might be more appropriate. In Section 2-5.1, we will investigate the determination of the systematic standard uncertainty b_X for various systematic error distribution assumptions.

The ultimate goal of uncertainty analysis is to find the range $(X_{best} \pm u_X)$ within which we think X_{true} falls. The standard uncertainty u_X is an estimate of the standard deviation of the parent population for the distribution of X considering the combination of all the errors affecting the measured value X. We do not know what the distribution will be when we combine the random errors with the systematic errors; however, a concept called the central limit theorem allows us to make a Gaussian (normal) assumption for X in many cases.

The central limit theorem states that if X is not dominated by a single error source but instead is affected by multiple, independent error sources, then the resulting distribution for X will be approximately normal [1]. The example below illustrates the power of the central limit theorem.

Consider a measurement X that is affected by multiple error sources δ_i as

$$X = X_{true} + \delta_1 + \delta_2 + \cdots + \delta_N$$

Let us take $X_{true} = 100$ and assume that the errors (δ_i) come from rectangular (uniform) distributions each having a range from -50 to $+50$, as shown in Figure 2.3.

We will look at two cases, one where X has two independent errors δ_1 and δ_2 and the other where X has eight independent error sources $\delta_1, \ldots, \delta_8$. We will assume that each error has the parent distribution given in Figure 2.3.

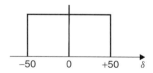

Figure 2.3   Distribution of errors for central limit theorem example.

For case 1, the possible values of X are given as

$$X = 100 + \delta_1 + \delta_2$$

In order to generate a value for X, we will randomly sample the uniform distribution in Figure 2.3 once for δ_1 and then again for δ_2 and add these errors to 100. To obtain an estimate of the parent distribution for X, we repeat this process many times. The resulting histogram for the X values is shown in Figure 2.4a

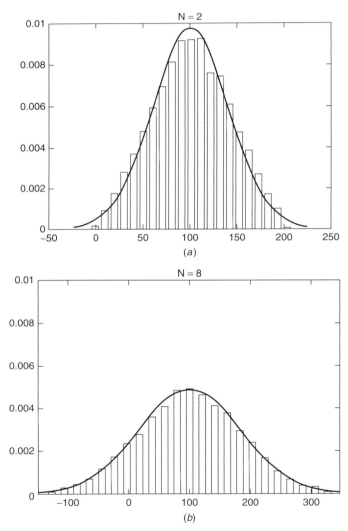

Figure 2.4   Distribution of values of measurement of variable X for (a) case 1 with two error sources and (b) case 2 with eight error sources.

for 10,000 iterations. Also shown in the figure is a Gaussian distribution that has a mean value of 100 and the same standard deviation as that for the 10,000 samples. It is seen that even though the two errors δ_1 and δ_2 came from rectangular distributions, the resulting (triangular) distribution for X appears closely Gaussian.

For case 2, we followed the same procedure with

$$X = 100 + \delta_1 + \delta_2 + \cdots + \delta_8$$

The histogram for X in this case is shown in Figure 2.4b with the associated Gaussian distribution. The resulting distribution for X is even closer to normal for this case. Of course, if we had only one error source for X that was rectangular, the resulting distribution would be rectangular since the central limit theorem would not apply.

Because of the importance of the Gaussian distribution in dealing with the effects of random errors and in the possible distributions for the measured variable through the central limit theorem, we will use it as an example as we cover the basic methodology.

In the following sections we discuss first the statistical characteristics of Gaussian parent distributions and then those of sample distributions that may contain relatively large or relatively small numbers of measurements. In addition, we consider how these statistical concepts developed for estimating the effects of random errors on the measurement of a variable can be extended to estimates of the effects of systematic errors. Finally, in Section 2-5 the combination of estimates of random and systematic uncertainties into an overall uncertainty in the measurement of a variable is considered.

2-2 GAUSSIAN DISTRIBUTION

For cases in which the variation in the measurements results from a combination of many small errors of equal magnitude with each of the errors being equally likely to be positive as negative, the smooth distribution of an infinite number of measurements coincides with the *Gaussian*, or *normal, distribution*. Also, a normal distribution will often be appropriate for a variable even if some of the error sources have distributions that are non-Gaussian (rectangular, triangular, etc.), a powerful result of the central limit theorem. The Gaussian distribution has been found to describe many real cases of experimental and instrument variability.

2-2.1 Mathematical Description

The equation for the Gaussian distribution is

$$f(X) = \frac{1}{\sigma\sqrt{2\pi}}e^{-(X-\mu)^2/2\sigma^2} \tag{2.1}$$

where $f(X)\, dX$ is the probability that a single measurement of X will lie between X and $X + dX$, μ is the distribution mean defined as

$$\mu = \lim_{N \to \infty} \frac{1}{N} \sum_{i=1}^{N} X_i \qquad (2.2)$$

and σ is the distribution standard deviation defined as

$$\sigma = \lim_{N \to \infty} \left[\frac{1}{N} \sum_{i=1}^{N} (X_i - \mu)^2 \right]^{1/2} \qquad (2.3)$$

The square of the standard deviation is known as the *variance* of the distribution.

A plot of Eq. (2.1) is shown in Figure 2.5 for two cases in which the mean μ is equal to 5.0 and the standard deviation is equal to 0.5 and 1.0, respectively. As the value of σ increases, the range of values of X expected also increases. Larger values of σ therefore correspond to cases in which the scatter in the X measurements is large and thus the range of potential random errors is large.

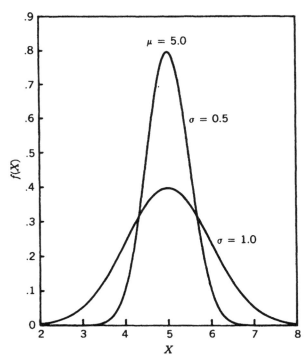

Figure 2.5 Plot of the Gaussian distribution showing the effect of different values of standard deviation.

Equation (2.1) is in a normalized form such that

$$\int_{-\infty}^{\infty} f(X)\, dX = 1.0 \tag{2.4}$$

This result makes intuitive sense, since we would expect that the probability of a measurement being between $\pm\infty$ must equal 1.

Now suppose that we wish to determine the probability that a single measurement from a Gaussian parent population will fall within a specified range, say $\pm\Delta X$, about the mean value μ. This can be expressed as

$$\mathrm{Prob}(\Delta X) = \int_{\mu-\Delta X}^{\mu+\Delta X} \frac{1}{\sigma\sqrt{2\pi}} e^{-(X-\mu)^2/2\sigma^2}\, dX \tag{2.5}$$

This integral cannot be evaluated in closed form, and if its value were tabulated for a range of ΔX values, there would have to be a table for every pair of (μ,σ) values. Rather than attempt to generate an infinite number of tables, it is more logical to normalize the integral so that only a single table is required.

If a normalized deviation from the mean value μ is defined as

$$\tau = \frac{X-\mu}{\sigma} \tag{2.6}$$

then Eq. (2.5) can be rewritten as

$$\mathrm{Prob}(\tau_1) = \frac{1}{\sqrt{2\pi}} \int_{-\tau_1}^{\tau_1} e^{-\tau^2/2}\, d\tau \tag{2.7}$$

where $\tau_1 = \Delta X/\sigma$.

The value of $\mathrm{Prob}(\tau_1)$ corresponds to the area under the Gaussian curve between $-\tau_1$ and $+\tau_1$, as shown in Figure 2.6. $\mathrm{Prob}(\tau)$ is called a *two-tailed*

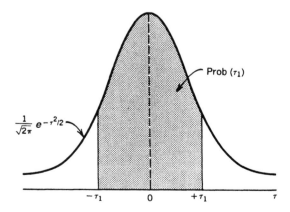

Figure 2.6 Graphic representation of the probability $\mathrm{Prob}(\tau_1)$.

probability since both the negative and positive tails of the distribution are included in the integration. Since the Gaussian distribution is symmetric, the probability of a measurement with a nondimensional deviation between 0 and τ or between $-\tau$ and 0 is $\frac{1}{2}\text{Prob}(\tau)$ and is called a *single-tailed probability*. Values of $\text{Prob}(\tau)$ for τ varying from 0.0 to 5.0 are presented in Table A.1 of Appendix A.

Example 2.1 A Gaussian distribution has a mean μ of 5.00 and a standard deviation σ of 1.00. Find the probability of a single measurement from this distribution being

   (a) Between 4.50 and 5.50
   (b) Between 4.50 and 5.75
   (c) Equal to or less than 6.50
   (d) Between 6.00 and 7.00

Solution

   (a) For $\mu = 5.0$ and $\sigma = 1.0$, we want $\text{Prob}(4.5 < X_i < 5.5)$. Let $X_1 = 4.50$ and $X_2 = 5.50$; then

$$\tau_1 = \frac{X_1 - \mu}{\sigma} = \frac{4.50 - 5.00}{1.00} = -0.50$$

$$\tau_2 = \frac{X_2 - \mu}{\sigma} = \frac{5.50 - 5.00}{1.00} = +0.50$$

Since $\tau_1 = \tau_2$, this is exactly what is tabulated in the two-tailed probability tables. Therefore, $\text{Prob} = \text{Prob}(0.5)$ (see Figure 2.7). From Table A.1,

$$\text{Prob}(\tau) = \text{Prob}(0.5) = 0.3829 \approx 38.3\%$$

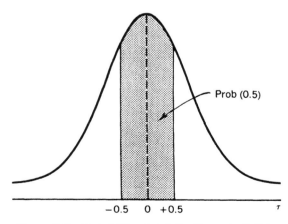

Figure 2.7 Probability sketch for Example 2.1(a).

(b) We want Prob($4.50 < X_i < 5.75$). Let $X_1 = 4.50$ and $X_2 = 5.75$; from part (a), $\tau_1 = -0.50$ and

$$\tau_2 = \frac{X_2 - \mu}{\sigma} = \frac{5.75 - 5.00}{1.00} = +0.75 \qquad \text{(see Figure 2.8)}$$

$$\text{Prob} = \tfrac{1}{2}\,\text{Prob}(0.5) + \tfrac{1}{2}\,\text{Prob}(0.75)$$

$$= \tfrac{1}{2}(0.3829) + \tfrac{1}{2}(0.5467) = 0.4648 \approx 46.5\%$$

(c) We want Prob($X_i \leq 6.50$); let

$$\tau_1 = \frac{X_i - \mu}{\sigma} = \frac{6.5 - 5.0}{1.0} = 1.50 \quad \text{(see Figure 2.9)}$$

$$\text{Prob} = 0.5 + \tfrac{1}{2}\,\text{Prob}(1.50) = 0.5 + \tfrac{1}{2}(0.8664)$$

$$= 0.9332 \approx 93.3\%$$

(d) We want Prob($6.00 < X_i < 7.00$). Let $X_1 = 6.00$ and $X_2 = 7.00$; then

$$\tau_1 = \frac{X_1 - \mu}{\sigma} = \frac{6.00 - 5.00}{1.00} = +1.0$$

$$\tau_2 = \frac{X_2 - \mu}{\sigma} = \frac{7.00 - 5.00}{1.00} = +2.0 \quad \text{(see Figure 2.10)}$$

$$\text{Prob} = \tfrac{1}{2}\,\text{Prob}(2.0) - \tfrac{1}{2}\,\text{Prob}(1.0)$$

$$= \tfrac{1}{2}(0.9545) - \tfrac{1}{2}(0.6827)$$

$$= 0.1359 \approx 13.6\%$$

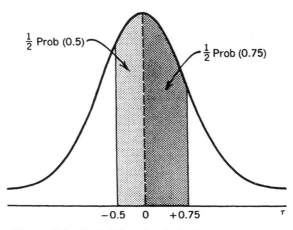

Figure 2.8   Probability sketch for Example 2.1(b).

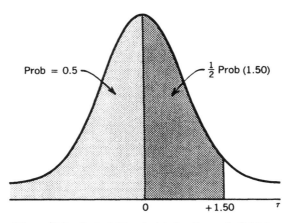

Figure 2.9   Probability sketch for Example 2.1(c).

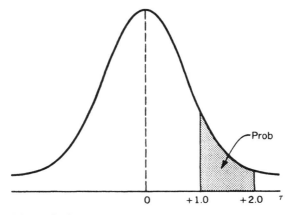

Figure 2.10   Probability sketch for Example 2.1(d).

2-2.2 Confidence Intervals in Gaussian Distribution

Based on what we have seen in Example 2.1 and on the probabilities tabulated in Table A.1, we can observe that 50% of the measurements from a Gaussian parent population are within $\pm 0.675\sigma$ of the mean, 68.3% are within $\pm 1.0\sigma$ of the mean, 95% are within $\pm 1.96\sigma$ of the mean, 99.7% are within $\pm 3.0\sigma$ of the mean, and 99.99% are within $\pm 4.0\sigma$ of the mean.

Consider a variable X that has a Gaussian parent population with a mean value μ and a standard deviation σ. If we were to make measurements of X, within what interval could we be 95% confident that any particular measurement X_i would fall?

Using Eq. (2.6), we can define the normalized deviation of X_i from μ as

$$\tau = \frac{X_i - \mu}{\sigma}$$

and we can obtain from Table A.1 that, for $\text{Prob}(\tau) = 0.95$, $\tau = 1.96$. This probability expression can be written as

$$\text{Prob}\left(-1.96 \leq \frac{X_i - \mu}{\sigma} \leq 1.96\right) = 0.95 \tag{2.8}$$

or, after multiplying the terms in the parentheses by σ and then adding μ to each term,

$$\text{Prob}(\mu - 1.96\sigma \leq X_i \leq \mu + 1.96\sigma) = 0.95 \tag{2.9}$$

Thus, knowing that 95% of the population lies within $\pm 1.96\sigma$ of the mean μ, we can be 95% confident that any particular measurement will fall within this $\pm 1.96\sigma$ interval about the mean. Stated another way, $+1.96\sigma$ and -1.96σ are the upper and lower bounds on the 95% confidence interval for the measurement of X. We can also speak of $\pm 1.96\sigma$ as the 95% confidence limits. This concept of a confidence interval is fundamental to uncertainty analysis in experimentation.

If we turned our point of view around $180°$, we might ask within what interval about a particular measurement X_i would we expect the mean value of the distribution to lie at a confidence level of 95%? Upon rearranging Eq. (2.8) to isolate μ rather than X_i, we find that

$$\text{Prob}(X_i - 1.96\sigma \leq \mu \leq X_i + 1.96\sigma) = 0.95 \tag{2.10}$$

Thus, as shown in Figure 2.11, we can be 95% confident that the mean μ of the distribution will fall within $\pm 1.96\sigma$ of the single measurement X_i. This conclusion follows since 95% of the measurements from the distribution will fall within $\pm 1.96\sigma$ of the mean μ. As we will see in the next section, the parent population mean value μ is usually not available. Therefore, this concept of a 95% confidence interval allows us to estimate the range that should contain μ.

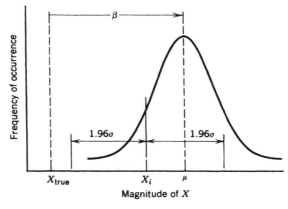

Figure 2.11   The 95% confidence interval about a single measurement from a Gaussian distribution.

2-3 SAMPLES FROM GAUSSIAN PARENT POPULATION

The preceding discussion has assumed that we were working with a Gaussian parent population which would be well described if we took an infinite number of measurements. However, it is impractical to expect anything approaching an infinite number of measurements in a realistic experimental situation. We must therefore consider the statistical properties of a sample population that consists of a finite number of measurements drawn from a parent distribution.

2-3.1 Statistical Parameters of Sample Population

The mean of the sample population is defined by

$$\overline{X} = \frac{1}{N} \sum_{i=1}^{N} X_i \tag{2.11}$$

where N is the number of individual measurements X_i. The standard deviation of the sample population is defined by

$$s_X = \left[\frac{1}{N-1} \sum_{i=1}^{N} (X_i - \overline{X})^2 \right]^{1/2} \tag{2.12}$$

where the $N-1$ occurs instead of N because the sample mean $\overline{X}$ is used instead of μ. This results in the loss of one degree of freedom since the same sample used to calculate s_X has already been used to calculate $\overline{X}$. Here, s_X is appropriately called the *sample standard deviation;* however, since we will be dealing with samples from realistic experiments for the remainder of this book, we will refer to s_X as the standard deviation. For clarity, σ will be called the *parent population standard deviation*.

A key point in the determination of the standard deviation of a variable is how well s describes the possible random variations of the variable. All the factors that influence the random variation must be present when the N measurements of the variable are taken. This statement may seem obvious, but with high-speed data acquisition, many data measurements can be taken in a very short period of time. If the random effects on the variable have a larger time frame, one set of computer-acquired measurements will not be sufficient to determine s, and the mean of these measurements will essentially be one measurement of the variable. Several sets of measurements or multiple independent measurements will have to be taken over the appropriate time frame. This concept of time frame for the determination of random uncertainty is stressed throughout the book and discussed further in Chapter 5.

Another statistical parameter of interest is the standard deviation associated with the sample mean $\overline{X}$. Suppose that five sets of $N = 50$ measurements are taken from a Gaussian parent population with mean μ and parent population standard deviation σ and that a mean value is calculated for each of the five sets using Eq. (2.11). We would certainly not expect the five mean values to be the same. In fact, the sample means are normally distributed [2] with mean μ and standard deviation

$$\sigma_{\overline{X}} = \frac{\sigma}{\sqrt{N}} \qquad (2.13)$$

The implications of this relationship have some practical importance. One way to decrease the random component of the uncertainty in a measured value is to take many measurements and average them. The inverse-square-root relationship of (2.13) indicates that this is a situation with rapidly diminishing returns—to reduce $\sigma_{\overline{X}}$ by a factor of 2, four times as many measurements are required, whereas 100 times as many measurements must be taken to reduce $\sigma_{\overline{X}}$ by a factor of 10.

Of course, the parent population standard deviation σ of the distribution is unknown, so in practice we must use the sample standard deviation of the mean, which is defined as

$$s_{\overline{X}} = \frac{s_X}{\sqrt{N}} \qquad (2.14)$$

where s_X is the standard deviation of the sample of N measurements as given by Eq. (2.12). For the remainder of this book we will refer to $s_{\overline{X}}$ as the standard deviation of the mean. As noted above, the N measurements must be independent and must cover the range of random variations of the variable in order for Eq. (2.14) to be an appropriate determination of the standard deviation of the mean.

2-3.2 Confidence Intervals in Sample Populations

As we saw in Section 2-2.2, for a Gaussian distribution with mean μ and parent population standard deviation σ, Eq. (2.10) was given as

$$\text{Prob}(X_i - 1.96\sigma \leq \mu \leq X_i + 1.96\sigma) = 0.95$$

Thus we can say with 95% confidence that the mean μ of the parent distribution is within $\pm 1.96\sigma$ of a single measurement from that distribution.

Recalling from the preceding section that the mean $\overline{X}$ of a sample population of size N drawn from this same Gaussian distribution is itself normally distributed with standard deviation $\sigma/\sqrt{N}$, we can write

$$\text{Prob}\left(\overline{X} - 1.96\frac{\sigma}{\sqrt{N}} \leq \mu \leq \overline{X} + 1.96\frac{\sigma}{\sqrt{N}}\right) = 0.95 \qquad (2.15)$$

Thus we can also say with 95% confidence that the mean μ of the parent distribution is within $\pm 1.96\sigma/\sqrt{N}$ of the sample mean $\overline{X}$ computed from N measurements. The width of the 95% confidence interval in Eq. (2.15) is narrower than the one in Eq. (2.10) by a factor of $1/\sqrt{N}$.

Once again, the problem we face in actual experimental situations is that we do not know the value of σ—what we have is s_X, the standard deviation of a finite sample of N measurements, which is only an estimate of the value of σ. Comparing Eqs. (2.12) and (2.3), we can see that the value of s_X approaches σ as the number of measurements N in the sample approaches infinity.

If we are still interested in determining a 95% confidence interval, we are then forced to follow the same approach as shown in Eq. (2.8) and to seek the value of t that satisfies

$$\text{Prob}\left(-t_{95} \leq \frac{X_i - \mu}{s_X} \leq t_{95}\right) = 0.95 \tag{2.16}$$

and

$$\text{Prob}\left(-t_{95} \leq \frac{\overline{X} - \mu}{s_X/\sqrt{N}} \leq t_{95}\right) = 0.95 \tag{2.17}$$

where t_{95} is no longer equal to 1.96 because s_X is only an estimate of σ based on a finite number of measurements N.

The variables $(X_i - \mu)/s_X$ and $(\overline{X} - \mu)/(s_X/\sqrt{N})$ are not normally distributed—that is, their behavior is not Gaussian. Rather, they follow the t distribution with $N - 1$ degrees of freedom ν and the values of t that satisfy (2.16) and (2.17) are functions of the size N of the sample population. Table A.2 in Appendix A presents tabular values of t as a function of ν for several different confidence levels.

From Table A.2 we see that the 95% confidence interval is wider for smaller N: $t_{95} = 2.57$ for $N = 6$ as opposed to $t_{95} = 2.04$ for $N = 31$, for instance. The 95% confidence level value of t approaches the Gaussian value of 1.96 as N approaches infinity.

Upon rearranging Eqs. (2.16) and (2.17) to isolate μ, the confidence interval expressions become

$$\text{Prob}(X_i - t_{95}s_X \leq \mu \leq X_i + t_{95}s_X) = 0.95 \tag{2.18}$$

and

$$\text{Prob}\left(\overline{X} - t_{95}\frac{s_X}{\sqrt{N}} \leq \mu \leq \overline{X} + t_{95}\frac{s_X}{\sqrt{N}}\right) = 0.95 \tag{2.19}$$

These expressions give the range around a single measurement or the mean of N measurements, respectively, where the parent population mean μ will lie with 95% confidence. For other confidence levels, the appropriate values of t would be used from Table A.2.

Example 2.2   A thermometer immersed in an insulated beaker of fluid is read independently by 24 students. The thermometer has a least scale division of 0.1°F. The mean of the sample of 24 measurements is $\overline{T} = 97.22°F$ and the standard deviation of the sample is $s_T = 0.085°F$, as calculated using Eq. (2.12).

(a) Determine the interval within which we expect, with 95% confidence (20:1 odds), the mean value μ_T of the parent population to fall.

(b) One additional temperature measurement of $T = 97.25°F$ is taken. Within what range about this measurement will the parent population mean μ_T fall with 95% confidence (20:1 odds)?

Solution

(a) From Eq. (2.19), the interval defined by $\overline{T} \pm t_{95}s_{\overline{T}}$ where

$$t_{95}s_{\overline{T}} = \frac{t_{95}s_T}{\sqrt{N}}$$

will include μ_T with 95% confidence.
For $N = 24$, the number of degrees of freedom ν is 23, and we find that

$$t_{95} = 2.069$$

from Table A.2. Using this value yields

$$t_{95}s_{\overline{T}} = \frac{(2.069)(0.085°F)}{\sqrt{24}} = 0.036°F$$

The interval given by

$$\overline{T} \pm t_{95}s_{\overline{T}} = 97.22°F \pm 0.04°F$$

should include μ_T with 95% confidence. Referring back to Figure 2.11, we note that μ_T is the *biased* mean value and would only correspond to the true temperature T_{true} if the systematic error (bias) in the temperature measurements were zero. Also note the commonsense approach used in rounding off the final number for the confidence interval. Only one significant figure is maintained in $t_{95}s_{\overline{T}}$, corresponding to the number of decimal places for $\overline{T}$.

(b) From Eq. (2.18) the interval defined by $T \pm t_{95}s_T$ will contain μ_T with 95% confidence. Using the information from the 24 previous measurements to obtain s_T, the degrees of freedom for t_{95} will be 23, yielding the same value for t_{95} as in part (a). The limit is then

$$t_{95}s_T = (2.069)(0.085°F) = 0.18°F$$

and the interval given by

$$T \pm t_{95}s_T = 97.25°F \pm 0.18°F$$

will contain μ_T with 95% confidence.

Note that the interval in part (b) is larger than that in part (a) because we are using only the single measurement as our "best estimate" of μ_T. This case is an example of using previous information about a measurement to estimate the standard deviation of the current reading. This topic is discussed further in Chapter 5.

Sometimes the previous information used to estimate s may come from multiple sets of measurements taken at different times. If it can be properly assumed that all of these measurements come from the same parent population with standard deviation σ, then a pooled standard deviation s_{pooled} can be used to determine the appropriate standard deviation to estimate σ [3]. Considering Example 2.2, if a group of eight students measured the temperature and determined the standard deviation s_1, and then separate groups of 6 and 10 students each did the same, yielding s_2 and s_3, respectively, then the pooled standard deviation for T would be

$$s_{T_{pooled}}^2 = \frac{(8-1)s_1^2 + (6-1)s_2^2 + (10-1)s_3^2}{(8-1) + (6-1) + (10-1)}$$

This expression can be generalized for any number of separate standard deviation determinations from the same parent population as

$$s_{pooled}^2 = \frac{(N_1-1)s_1^2 + (N_2-1)s_2^2 + \cdots + (N_k-1)s_k^2}{(N_1-1) + (N_2-1) + \cdots + (N_k-1)} \tag{2.20}$$

The use of an "odds" statement (as in Example 2.2) rather than a "percent confidence" statement is sometimes encountered when uncertainties are quoted. Although 19:1 odds (19 chances out of 20) strictly corresponds to a fractional value of 0.95, in practice it is commonplace for 20:1 odds and 95% confidence to be used interchangeably. The same holds true for 2:1 odds and 68% confidence, although two chances out of three corresponds to a fractional value of 0.667.

2-3.3 Tolerance and Prediction Intervals in Sample Populations

In addition to the confidence interval for a sample population, there are two other useful statistical intervals, the tolerance interval and the prediction interval [3, 4]. The tolerance interval gives information about the parent population that the samples came from, and the prediction interval is used to obtain the range within which future measurements from the same parent distribution will fall.

For the tolerance interval, first consider the Gaussian parent population. We know that the interval that contains 95% of the parent population is $\mu \pm 1.96\sigma$.

However, when we are dealing with samples and the statistical quantities $\overline{X}$ and s_X, the determination of the range that contains a portion of the parent population is not as straightforward. What we can estimate is the range that has a certain probability of containing a specified percentage of the parent population.

Consider a case where a sample of six measurements is used to calculate $\overline{X}$ and s_X. The interval that would contain 90% of the parent population with 95% confidence is then determined as

$$\overline{X} \pm [c_{T95(90)}(6)]s_X \tag{2.21}$$

where the c_T factors are given as a function of the number of readings N in the sample in Table A.3 in Appendix A. From the table we obtain a tolerance interval factor of 3.712. In Table A.3, tolerance interval factors are given for ranges containing 90%, 95%, or 99% of the parent population with 90%, 95%, or 99% confidence. Note that as $N \to \infty$, the concept of a confidence level for the tolerance interval does not apply because the tolerance interval approaches the Gaussian interval.

Example 2.3 The ultimate strength of an unknown alloy is determined by pulling specimens in a tensile test machine. Seven specimens are tested with ultimate strengths of (all in psi)

65,340	67,702
68,188	66,954
67,723	65,945
66,453	

Find the tolerance interval that contains

(a) 99% of the parent population with 90% confidence
(b) 99% of the parent population with 95% confidence
(c) 99% of the parent population with 99% confidence

Solution

(a) The tolerance interval that contains 99% of the parent population with 90% confidence is

$$\overline{X} \pm [c_{T90(99)}(7)]s_X$$

where

$$\overline{X} = 66,901 \text{ psi}$$

and

$$s_X = 1043 \text{ psi}$$

From Table A.3, the tolerance interval factor $c_{T90(99)}(7)$ is 4.521. The interval is then

$$66,901 \text{ psi} \pm (4.521)(1043) \text{ psi}$$

or

$$66{,}901 \text{ psi} \pm 4715 \text{ psi}$$

Using a commonsense approach to rounding, the interval is 66,900 psi $\pm$ 4700 psi.

(b) The tolerance interval that contains 99% of the parent population with 95% confidence is

$$\overline{X} \pm [c_{T95(99)}(7)]s_X$$

where, from Table A.3, $c_{T95(99)}(7) = 5.248$. The interval is then

$$66{,}901 \text{psi} \pm (5.248)(1043) \text{ psi}$$

or

$$66{,}900 \text{ psi} \pm 5500 \text{ psi}$$

(c) The tolerance interval that contains 99% of the parent population with 99% confidence is

$$\overline{X} \pm [c_{T99(99)}(7)]s_X$$

where, from Table A.3, $c_{T99(99)}(7) = 7.187$. The interval is then

$$66{,}901 \text{ psi} \pm (7.187)(1043) \text{ psi}$$

or

$$66{,}900 \text{ psi} \pm 7500 \text{ psi}$$

If we wish to estimate the interval that will contain future measurements from a parent population, a prediction interval is required. The prediction interval gives us the range that should contain all of 1, 2, 5, 10, ... future measurements from the population with a given confidence level. Consider again a case where a sample of six measurements is used to calculate $\overline{X}$ and s_X. The interval that would contain one additional measurement with 95% confidence is then determined as

$$\overline{X} \pm [c_{p,1}(6)]s_X \qquad (2.22)$$

where the c_p factors for 95% confidence are given as a function of N in Table A.4 in Appendix A. From the table we obtain a prediction interval factor of 2.78. Note that for the prediction interval, as $N \to \infty$, only the range that will contain one additional measurement approaches the Gaussian interval. Even for an infinite number of measurements for N, the predicted range that will contain all of 2, 5, 10, or 20 additional measurements is larger than the Gaussian interval.

Example 2.4   For the ultimate strength test described in Example 2.3, find the prediction interval that would contain (with 95% confidence)

(a) One additional result

(b) Both of 2 additional test results

(c) All of 10 additional test results

Solution

(a) The mean of the ultimate strength measurements for Example 2.3 is $\overline{X} = 66{,}901$ psi with $s_X = 1043$ psi. The prediction interval that would contain one additional test result at 95% confidence is

$$\overline{X} \pm [c_{p,1}(7)]s_X$$

where, from Table A.4, $c_{p,1}(7) = 2.62$. The prediction interval is then

$$66{,}901 \text{ psi} \pm (2.62)(1043) \text{ psi}$$

or

$$66{,}900 \text{ psi} \pm 2700 \text{ psi}$$

(b) The prediction interval that would contain both of two additional test results with 95% confidence is

$$\overline{X} \pm [c_{p,2}(7)]s_X$$

where, from Table A.4, $c_{p,2}(7) = 3.11$. The prediction interval is then

$$66{,}901 \text{ psi} \pm (3.11)(1043) \text{ psi}$$

or

$$66{,}900 \text{ psi} \pm 3200 \text{ psi}$$

(c) For the range that would contain all of 10 additional test results, the prediction interval is
$$\overline{X} \pm [c_{p,10}(7)]s_X$$

where $c_{p,10}(7) = 4.26$ from Table A.4. The prediction interval is then

$$66{,}901 \text{ psi} \pm (4.26)(1043)\text{psi}$$

or

$$66{,}900 \text{ psi} \pm 4400 \text{ psi}$$

2-4 STATISTICAL REJECTION OF OUTLIERS FROM A SAMPLE

When a sample of N measurements of a particular variable is examined, some measurements may appear to be significantly out of line with the others. Such points are often called *outliers*. If obvious, verifiable problems with the experiment when such points were taken are identified, they can be discarded. However, the more common situation is when there is no obvious or verifiable reason for the large deviation of these data points.

If one is making measurements from a Gaussian parent distribution, there is no guarantee that the second measurement will not be six standard deviations from the mean, thus essentially ruining a small sample estimate of the standard deviation. Since one wants a good estimate of the standard deviation—often with resources dictating a small number of measurements—then use of an objective statistical criterion to identify points that might be considered for rejection offers a useful engineering approach. There is no other justifiable way to "throw away" data points.

One method that has achieved relatively wide acceptance is *Chauvenet's criterion*, which defines an acceptable scatter (from a probability viewpoint) around the mean value from a given sample of N measurements from the same parent population. The criterion specifies that all points should be retained that fall within a band around the mean value that corresponds to a probability of $1 - 1/(2N)$. Stated differently, we can say that data points can be considered for rejection if the probability of obtaining their deviation from the mean value is less than $1/(2N)$. This criterion is shown schematically in Figure 2.12.

As an illustration of this criterion, consider a test in which six measurements are taken at a "constant" test condition. According to Chauvenet's criterion, all of the points that fall within a probability band around the mean of $1 - \frac{1}{12}$, or 0.917, must be retained. This probability can be related to a definite deviation away from the mean value by using the Gaussian probabilities in Table A.1. (Note that Chauvenet's criterion does not use t-distribution probabilities, even when N is small.) For a probability of 0.917, the nondimensional deviation τ equals 1.73 on interpolating in the table. Thus

$$\tau = \left| \frac{X_i - \overline{X}}{s_X} \right| = \left| \frac{\Delta X_{\max}}{s_X} \right| \tag{2.23}$$

where $\Delta X_{\max}$ is the maximum deviation allowed away from the mean value $\overline{X}$ of the six measurements and s_X is the standard deviation of the sample of six points. Therefore, all measurements that deviate from the mean by more than

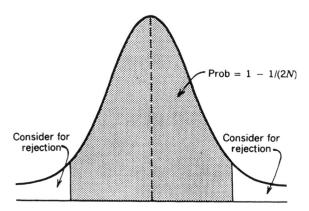

Figure 2.12 Graphic representation of Chauvenet's criterion.

Table 2.1   Chauvenet's Criterion for Rejecting a Reading

Number of Readings, N	Ratio of Maximum Acceptable Deviation to Standard Deviation, $|(\Delta X_{max}/s_X)|$
3	1.38
4	1.54
5	1.65
6	1.73
7	1.80
8	1.87
9	1.91
10	1.96
15	2.13
20	2.24
25	2.33
50	2.57
100	2.81
300	3.14
500	3.29
1000	3.48

$(1.73)s_X$ can be considered for rejection. Then a new mean value and a new standard deviation are calculated from the measurements that remain. No further application of the criterion to the sample is recommended—Chauvenet's criterion should be applied only once for a given sample of measurements.

This statistical method for identifying measurements that might be considered for rejection depends on the number of measurements, N, in the sample. Table 2.1 gives the maximum acceptable deviations for various sizes. Other values for N can be determined using the Gaussian probabilities in Table A.1.

Example 2.5   A laboratory has determined the ultimate strength of five specimens of an alloy on their tensile testing machine, and the values (in psi) are reported as follows:

i	X_i	$|(X_i - \overline{X})/s_X|$
1	65,300	0.26
2	68,000	0.52
3	67,700	0.49
4	43,600	1.78
5	67,900	0.51

Calculation of $\overline{X}$ and s_X gives

$$\overline{X} = 62,500 \text{ psi} \qquad s_X = 10,624 \text{ psi}$$

Having noticed that point 4 appears out of line with the others, we apply Chauvenet's criterion: For $N = 5$,

$$\frac{\Delta X_{max}}{s_X} = 1.65$$

We see from the third column that point 4 is indeed outside the range defined as statistically acceptable by the criterion.

If point 4 is rejected, then

$$\overline{X} = 67{,}225 \text{ psi} \qquad s_X = 1289 \text{ psi}$$

Comparing these with the values calculated previously, $\overline{X}$ is increased by only 8%, whereas s_X is decreased by a factor of 8.

Example 2.6 Dynamic pressure measurements (in in. H_2O) were made by 10 different students from a pressure gauge connected to a Pitot probe exposed to "constant"-flow conditions. The 10 measurements were as follows:

i	X_i	$|(X_i - \overline{X})/s_X|$
1	1.15	0.07
2	1.14	0.00
3	1.01	0.87
4	1.10	0.27
5	1.11	0.20
6	1.09	0.33
7	1.10	0.27
8	1.10	0.27
9	1.55	2.73
10	1.09	0.33

Just looking down the X_i column, the ninth measurement appears to be an outlier, but we have no basis on which to reject it except by statistical analysis. We apply Chauvenet's criterion as follows:

$$\overline{X} = 1.14 \qquad s_X = 0.15$$

For $N = 10$, $\Delta X_{max}/s_X = 1.96$ from Table 2.1. If reading 9 is rejected, recalculation of $\overline{X}$ and s_X yields

$$\overline{X} = 1.10 \qquad s_X = 0.04$$

Note that rejection of the probable outlier changed the mean value by only about 4%, but s_X, the estimate of the standard deviation, is only about one-fourth of the value calculated originally.

2-5 UNCERTAINTY OF A MEASURED VARIABLE

In the preceding sections, we have seen that quantification of the standard uncertainty due to random errors in the measurements of a variable can be achieved by using s, where s is the standard deviation of the sample of N measurements. We found that, for a Gaussian distribution, the confidence interval $t_{\%}s$ around the single measurement of a variable contains the parent population mean μ, where $t_{\%}$ is the value from the t distribution that gives a particular percent confidence level. We also found that the interval $t_{\%}s_{\overline{X}}$ around the variable mean value will contain μ at a given percent level of confidence.

But in uncertainty analysis we are not just concerned with an interval that contains μ. We want to know an interval around the measured variable that will contain the true value of the variable, X_{true}, at a given level of confidence. In order to find this interval for X_{true}, we must quantify the systematic standard uncertainty b (the uncertainty component that arises from the effect of the systematic errors) and must then combine the random and systematic standard uncertainties to obtain the combined standard uncertainty u. The question then becomes how to determine the expanded uncertainty estimate U_X so that the interval $X_{\text{best}} \pm U_X$ will contain the true value of X at a given level of confidence. These topics are covered next.

2-5.1 Systematic Standard Uncertainty Estimation

Consider the situation shown in Figure 2.13, where the distribution of a sample of measurements of a variable X is shown. The random standard uncertainty has been considered, but what about the systematic contribution to the uncertainty?

Sometimes, as in a calibration, we have errors that are random in the instrument calibration process but which then are frozen or "fossilized" when the instrument is used to make measurements. We can use the calibration data to calculate the standard deviation and thus the systematic standard uncertainty b to account for these errors that become fossilized or fixed. This subject is discussed further in Chapter 5.

The more typical situation for systematic errors is when there is no sample of N measurements to quantify b. All that we know is that there is a systematic error β as shown in Figure 2.13, but we never know what it is because we do not know the true value. We know that β is the fixed error that remains after all calibration corrections are made and that it is the sum of all the significant systematic errors. For example, for three significant error sources

$$\beta = \beta_1 + \beta_2 + \beta_3 \qquad (2.24)$$

What we need are the standard uncertainties b_1, b_2, and b_3 that quantify the effects of these errors.

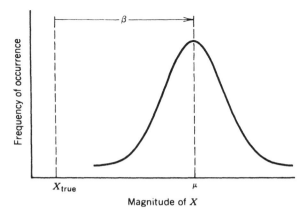

Figure 2.13   Systematic error in a sample of measurements of a variable X.

A useful approach to estimating the magnitude of a systematic error is to assume that the systematic error for a given case is a single realization drawn from some statistical parent distribution of possible systematic errors, as shown in Figure 2.14. For example, suppose a thermistor manufacturer specifies that 95% of the samples of a given model are within $\pm 1.0°$C of a reference resistance–temperature $(R-T)$ calibration curve supplied with the thermistors. One might assume that the systematic errors (the differences between the actual, but unknown, $R-T$ curves of the various thermistors and the reference curve) belong to a Gaussian parent distribution with a standard deviation $b = 0.5°$C. Then the interval defined by $\pm 2b = \pm 1.0°$C would include about 95% of the possible systematic errors that could be realized from the parent distribution, where here we are rounding the factor 1.96 in Eq. (2.9) to 2.

In another case, we may be reasonably sure that the systematic error is between the limits $\pm A$ but have no basis to say that it is closer to $+A$, $-A$, or 0. In this case, a rectangular, or uniform, distribution could be assumed for β as shown in Figure 2.15. The systematic standard uncertainty estimate would then be $b = A/\sqrt{3}$, which is the standard deviation of the rectangular distribution.

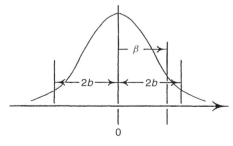

Figure 2.14   Distribution of possible systematic errors β and the interval with a 95% level of confidence of containing a particular β value.

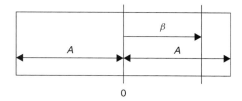

Figure 2.15   Uniform distribution of possible systematic errors β.

A variation on the above case would be if we thought the systematic error was between the limits $\pm A$ but suspected that it was more likely to be zero than either $+A$ or $-A$. In this case, a triangular distribution would be appropriate as in Figure 2.16, and the systematic standard uncertainty would be $b = A/\sqrt{6}$ (the standard deviation for the triangular distribution).

The key point is that a standard deviation estimate must be made for the distributions for each of the systematic error sources identified as being significant in the measurement of a variable. It is up to the analyst to use the best information and judgment possible to make the estimate.

2-5.2 Overall Uncertainty of a Measured Variable

The overall uncertainty of a measured variable X is the interval around the best value of X within which we expect the true value X_{true} to lie with a given confidence level. To obtain the overall uncertainty, we must appropriately combine the random and systematic standard uncertainty estimates.

The ISO guide [1] states that the method for combining uncertainty estimates is to add the variances (or the squares of the standard deviations) for the estimates. As we discussed in the preceding section, the standard deviation estimate for the systematic uncertainty for error source k is b_k, or our best estimate of the standard deviation of the possible parent population for the systematic error β_k.

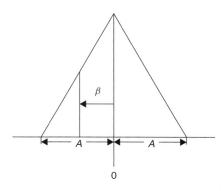

Figure 2.16   Triangular distribution of possible systematic errors β.

The standard deviation estimate for the random uncertainty is s_X from either Eq. (2.12) or (2.14) as appropriate depending on whether the uncertainty interval is centered on X or $\overline{X}$. Note that the b_k's have the same values regardless of whether X is a single reading or a mean value. The averaging process does not affect the systematic uncertainties.

Following the methodology of the ISO guide [1], the *combined standard uncertainty* u_c for variable X is

$$u_c^2 = s_X^2 + \sum_{k=1}^{M} b_k^2 \qquad (2.25)$$

where M is the number of significant elemental systematic error sources.

In order to associate a level of confidence with the uncertainty for the variable, the ISO guide recommends a coverage factor such that

$$U_\% = k_\% u_c \qquad (2.26)$$

where $U_\%$ is the overall or *expanded uncertainty* at a given percent level of confidence. The question now is: What is the appropriate coverage factor to use to obtain the overall uncertainty? As discussed previously, the central limit theorem tells us that the distribution for the total errors δ for the variable will usually approach Gaussian, where the factor u_c is in a sense an estimate of the standard deviation of this total error distribution. For this reason, the ISO guide recommends using the values from the t distribution (Table A.2) to obtain $k_\%$, so that

$$U_\% = t_\% u_c \qquad (2.27)$$

The $\pm U_\%$ band around the variable (X or $\overline{X}$) will contain the true value of the variable with the given percent level of confidence.

A number of degrees of freedom is needed in order to select the t value from Table A.2 for a given level of confidence. The effective number of degrees of freedom ν_X for determining the t value is approximated by the Welch–Satterthwaite formula [1] as

$$\nu_X = \frac{\left(s_X^2 + \sum_{k=1}^{M} b_k^2\right)^2}{s_X^4/\nu_{s_X} + \sum_{k=1}^{M} b_k^4/\nu_{b_k}} \qquad (2.28)$$

where ν_{s_X} is the number of degrees of freedom associated with s_X and ν_{b_k} is the number of degrees of freedom associated with b_k. As discussed previously,

$$\nu_{s_X} = N - 1 \qquad (2.29)$$

The ISO guide [1] recommends an estimate of ν_{b_k} as

$$\nu_{b_k} \approx \frac{1}{2}\left(\frac{\Delta b_k}{b_k}\right)^{-2} \qquad (2.30)$$

where the quantity in parentheses is the relative uncertainty of b_k.

Example 2.7 Estimate the number of degrees of freedom for a systematic uncertainty estimate b_k if the relative uncertainty in the value of $2b_k$ is $\pm 25\%$ (Figure 2.17).

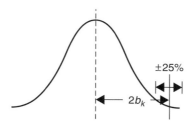

Figure 2.17   Relative uncertainty in systematic uncertainty estimate.

Solution   If the relative uncertainty in $2b_k$ is

$$\frac{\Delta 2b_k}{2b_k} = \frac{\Delta b_k}{b_k} = 0.25$$

the approximate number of degrees of freedom is

$$\nu_{b_k} \approx \tfrac{1}{2}(0.25)^{-2} = 8$$

Note that as our estimate of the relative uncertainty in b_k goes up, the number of degrees of freedom goes down, and alternatively, as we become more certain of our estimate of b_k, the number of degrees of freedom goes up.

2-5.3 Large-Sample Uncertainty of a Measured Variable

As discussed in the previous section, an estimate of the degrees of freedom ν_X of a variable is needed in order to determine the expanded uncertainty. Using the approximation in Eq. (2.28), we see that the degrees of freedom for the variable is dependent on the degrees of freedom for the random standard uncertainty and the effective degrees of freedom for the systematic standard uncertainties. What we find is that unless one error source dominates the uncertainty, the degrees of freedom for the variable will be larger than the individual degrees of freedom of the random and systematic components of the uncertainty.

Consider a case where there are three systematic error sources and the standard uncertainties s_X, b_1, b_2, and b_3 are all equal and all have the same value d for degrees of freedom. For this case, Eq. (2.28) reduces to $\nu_X = 4d$. So, for example, if each standard uncertainty has a degrees of freedom of 8, the degrees of freedom for X will be 32.

A consequence of using the effective degrees of freedom from Eq. (2.28) for determining the t value to calculate the expanded uncertainty [Eq. (2.27)] is that for most real engineering and scientific experiments the degrees of freedom will be large enough to consider the t value equal to a constant. This constant will be approximately equal to the Gaussian value for a given level of confidence (i.e., 2 for 95% or 2.6 for 99%). To illustrate this statement, we used a Monte Carlo simulation to study the behavior of the expanded uncertainty of X as a function of the degrees of freedom [5]. We considered a 95% level of confidence.

To start the simulation, we assumed that the central limit theorem applied so that the errors for the variable X were from a Gaussian parent population with mean X_{true} and standard deviation σ. Then a numerical experiment was run using this parent population. For $\nu = 8$, for example, the results were calculated as follows. Nine measurements were drawn randomly from the parent population specified, and a sample standard deviation was calculated using Eq. (2.12). This standard deviation was taken as the combined standard uncertainty u for X. A check was then made to see if the $\pm t_{95}u$ and $\pm 2u$ intervals centered around the next (tenth) randomly chosen measurement covered X_{true}, and the appropriate counter was incremented when the check was positive. This procedure was followed for 100,000 trials, and the percent coverage was the percentage of the 100,000 trials in which X_{true} was covered. This entire procedure was followed for ν from 2 to 29, and the results are shown in Figure 2.18. The expanded uncertainties using the factors t_{95} and the large-sample constant approximation of 2 converge quickly even for relatively small degrees of freedom for X. Since the degrees of freedom is not just the number of measurements minus 1 ($N - 1$) from the random standard uncertainty but instead is influenced by the systematic uncertainty component also, the large-sample approximation applies in most cases. Similar results are seen at other levels of confidence.

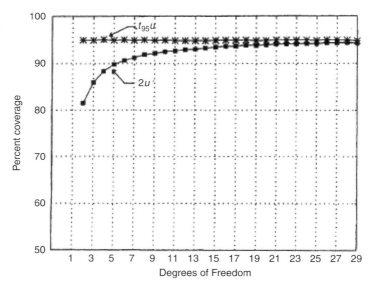

Figure 2.18 Percent coverage of true value X_{true} for a measurement X with various degrees of freedom.

The resulting large-sample approximation for a 95% level of confidence ($t_{95} \approx 2$) is determined from Eqs. (2.25) and (2.27) as

$$U_{95} = 2 \left(s_X^2 + \sum_{k=1}^{M} b_k^2 \right)^{1/2} \qquad (2.31)$$

The true value of the variable will then be within the limits

$$X - U_{95} \leq X_{\text{true}} \leq X + U_{95} \qquad (2.32)$$

with a 95% level of confidence where X is either X or $\overline{X}$ as appropriate.

A similar expression can be developed for a large-sample 99% level of confidence estimate of the uncertainty of a variable. For large degrees of freedom, $t_{99} \approx 2.6$, and

$$U_{99} = 2.6 \left[s_X^2 + \sum_{k=1}^{M} b_k^2 \right]^{1/2} \qquad (2.33)$$

2-5.4 Uncertainty of Measured Variable by Monte Carlo Method

As we will discuss in Chapter 3, the Monte Carlo Method (MCM) is a powerful tool for doing uncertainty analysis. The basic methodology for performing a Monte Carlo simulation was described in Section 2-5.3.

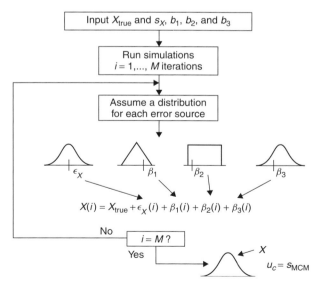

Figure 2.19　Monte Carlo Method for determining the combined standard uncertainty of a variable.

The application of the MCM for determining the combined standard uncertainty of a variable is shown in Figure 2.19. For each error source, an appropriate probability distribution function is chosen. In Figure 2.19, a Gaussian distribution is chosen for the random error and a triangular, rectangular, and Gaussian distribution, respectively, are chosen for each of the systematic errors where the standard deviations are taken as s_X and b_1, b_2, and b_3. A random sample is taken from each distribution and these are added to X_{true} to get the value of X for the ith iteration $X(i)$. The process is repeated until a converged value is obtained for the standard deviation, s_{MCM}, of the resulting $X(i)$ values. The converged standard deviation is then the estimate of the combined standard uncertainty u_c for X. Assuming that the central limit theorem and the large-sample approximations apply, the expanded uncertainty for X would then be the product of s_{MCM} and the large-sample t value for a given level of confidence.

2-6 SUMMARY

In this chapter we have reviewed the basic statistical concepts necessary for uncertainty analysis. When we discuss or report the value of a measured quantity, we should also include the estimate of the overall or expanded uncertainty at the appropriate level of confidence with the associated estimates of the systematic and random standard uncertainties. In later chapters we take a detailed look at how various systematic and random standard uncertainties in measured variables combine into an overall uncertainty in an experimental result, and we also consider further ways in which actual estimates of the systematic and random components of uncertainty can be made.

REFERENCES

1. International Organization for Standardization (ISO), *Guide to the Expression of Uncertainty in Measurement*, ISO, Geneva, 1993.
2. Bowker, A. H., and Lieberman, G. J., *Engineering Statistics*, Prentice-Hall, Upper Saddle River, NJ, 1959.
3. Montgomery, D. C., and Runger, G. C., *Applied Statistics and Probability for Engineers*, 4th ed., Wiley, New York, 2007.
4. Hahn, G., "Understanding Statistical Intervals," *Industrial Engineering*, Dec. 1970, pp. 45–48.
5. Coleman, H. W., and Steele, W. G., "Engineering Application of Experimental Uncertainty Analysis," *AIAA Journal*, Vol. 33, No. 10, 1995, pp. 1888–1896.

PROBLEMS

2.1 A Gaussian distribution has a mean μ of 6.0 and a parent population standard deviation σ of 2.0. Find the probability of a single reading from this distribution being **(a)** between 5.0 and 7.0; **(b)** between 7.0 and 9.0; **(c)** between 5.4 and 7.8; **(d)** equal to or less than 6.4.

2.2 An electric motor manufacturer has found after many measurements that the mean value of the efficiency of his motors is 0.94. Variations in the motors and random errors in the test equipment and test procedure cause a parent population standard deviation of 0.05 in the efficiency determination. With this same test arrangement, what is the probability that a motor's efficiency may be determined to be 1.00 or greater?

2.3 The parent population standard deviation for using a certain radar speed detector is 0.6 mph for vehicle speeds around 35 mph. If the police use this detector, what is the probability that you would be clocked at below the speed limit if you were going 36 mph in a 35-mph speed zone?

2.4 A military cargo parachute has an automatic opening device set to switch at 200 m above the ground. The manufacturer of the device has specified that the parent population standard deviation is 80 m about the nominal altitude setting. If the parachute must open 50 m above the ground to prevent damage to the equipment, what percentage of the loads will be damaged?

2.5 An accurate ohmmeter is used to check the resistance of a large number of 2-Ω resistors. Fifty percent of the readings fall between 1.8 and 2.2 Ω. What is an estimate of the parent population standard deviation for the resistance value of the resistors, and within what range would the resistance of 95% of the resistors fall?

2.6 A pressure measurement system is known to have a parent population standard deviation of 2 psi. If a single pressure reading of 60 psi is read, within what range (at 95% confidence) would we expect the mean μ to fall for a large number of readings at this steady condition?

2.7 The value of a variable is measured nine times, and the random standard uncertainty of the mean is calculated to be 5%. How many measurements would you estimate would be necessary to find the mean with a random standard uncertainty of 2.5%?

2.8 The standard deviation s for a sample of four readings has been calculated to be 1.00. What are the values of the random standard uncertainty of the readings and of the random standard uncertainty of the mean of the readings?

2.9 A thermocouple is placed in a constant-temperature medium. Successive readings (in $^\circ$F) of 108.5, 107.6, 109.2, 108.3, 109.0, 108.2, and 107.8 are made. Within what range would you estimate (at 20:1 odds) the next reading will fall? Within what 95% confidence range would you estimate the parent population mean (μ) to lie? Estimate at 95% confidence the range of values for temperatures in which you would expect 95% of all possible measured values for this constant-temperature condition to lie. Estimate the range for 99% of all possible measured temperatures at 95% confidence.

2.10 The ACME Co. produces model 12B roadrunner inhibitors which are supposed to have a length of 2.00 m. You have been hired to determine the minimum number of samples (N) necessary for averaging if the plant manager must certify that the sample mean length ($\overline{X}$) is within 0.5% of the parent population mean (μ) with 95% confidence. You have been permitted to take length readings from nine randomly chosen model 12B's, and the results (in meters) were 2.00, 1.99, 2.01, 1.98, 1.97, 1.99, 1.99, 2.00, and 2.01. Estimate N and state all of your assumptions.

2.11 The spring constant of a certain spring (in lb/in.) has been determined multiple times as 10.36, 10.43, 10.41, 10.48, 10.39, 10.46, and 10.42. What is the best estimate (with 95% confidence) of the range that contains the parent population mean of the spring constant?

2.12 How would the range in problem 2.11 vary if the confidence level were changed to **(a)** 90%; **(b)** 99%; **(c)** 99.5%; **(d)** 99.9%?

2.13 For the data in problem 2.11, within what range would we expect all of 5 additional measurements to fall? What range would contain all of 20 additional measurements?

2.14 If 30 data measurements are taken, according to Chauvenet's criterion, what is the ratio of the maximum acceptable deviation to the standard deviation?

2.15 Ten different people measured the diameter of a circular plate and obtained results (in cm) of 6.57, 6.60, 6.51, 6.61, 6.59, 6.65, 6.61, 6.63, 6.62, and 6.64. Compare the sample mean and standard deviation before and after application of Chauvenet's criterion.

2.16 Twelve values of the calibration constant of an oxygen bomb calorimeter were determined (in cal/°F): 1385, 1381, 1376, 1393, 1387, 1400, 1391, 1384, 1394, 1387, 1343, and 1382. Using these data, determine the best estimate of the mean calibration constant with 95% confidence.

2.17 In a metal-casting facility, castings of specific types are weighed at random to check for consistency. The random standard uncertainty associated with the weight of a specific type of casting (considering material variation and instrument readability) is ±5 lb and the systematic standard uncertainty associated with the scale is ±5 lb. If the average weight of a casting is 500 lb, within what range would 95% of the weight measurements fall with 95% confidence? Within what range does the "true" average weight fall? If the scale were replaced with an identical model, within what range would we expect 95% of the weight measurements to fall with 95% confidence?

2.18 For the test in problem 2.11, it is estimated that the systematic standard uncertainty for the spring constant measurement is 0.05 lb/in. However, this systematic uncertainty estimate could be off by as much as ±20%. Estimate the range for the true value of the spring constant at 95% confidence. At 90% confidence. At 99% confidence.

3

UNCERTAINTY IN A RESULT DETERMINED FROM MULTIPLE VARIABLES

In the previous chapter, we presented the basic statistics for understanding uncertainty and the methods to quantify the random and systematic standard uncertainties for a single variable. In most cases, the result of our experiment or simulation will depend on several variables through a data reduction equation (DRE) or a simulation solution.

An example of an experimental result is the determination of the drag coefficient for a model in a wind tunnel test where drag force F_D, fluid density ρ, velocity V, and model frontal area A are measured or evaluated and the drag coefficient C_D is calculated as

$$C_D = \frac{2F_D}{\rho V^2 A} \tag{3.1}$$

A simple simulation solution would be the prediction of the temperature distribution T in a plane wall of thickness L by solving the equation

$$k\frac{d^2T}{dX^2} = 0 \tag{3.2}$$

with boundary conditions

$$T(X = 0) = T_1 \tag{3.3}$$

and

$$T(X = L) = T_2 \tag{3.4}$$

where k is the thermal conductivity. Both the experiment and simulation results will have uncertainty because of the uncertainty in the input variables. For the drag coefficient, force, density, velocity, and area all have uncertainties that will propagate through the DRE [Eq. (3.1)], resulting in an uncertainty for C_D. For the simulation, the boundary condition temperatures, the thermal conductivity, and the wall thickness will have uncertainties that yield an uncertainty in the predicted temperature at a position in the wall.

In this chapter, we will consider methods to propagate the uncertainties of the different variables into the determined result. Two propagation methods will be presented—the Taylor Series Method (TSM) and the Monte Carlo Method (MCM).

3-1 TAYLOR SERIES METHOD FOR PROPAPATION OF UNCERTAINTIES

The derivation of the Taylor Series Method (TSM) for propagating the uncertainties of the variables into the determined result is given in Appendix B. In this section, we will give the TSM propagation equation, first for a result that is a function of two variables and then for the more general case. The expansion of the combined standard uncertainty of the result to an uncertainty at a given level of confidence will then be discussed. A large-sample approximation is presented for cases where the degrees of freedom of the result are large. The section concludes with an example and a discussion of numerical techniques for doing the TSM propagation.

3-1.1 TSM for Function of Multiple Variables

For the case where the result r is a function of two variables x and y,

$$r = f(x, y) \tag{3.5}$$

the combined standard uncertainty of the result, u_r, is given by

$$u_r^2 = \left(\frac{\partial r}{\partial x}\right)^2 b_x^2 + \left(\frac{\partial r}{\partial y}\right)^2 b_y^2 + \left(\begin{array}{c} \text{systematic error} \\ \text{correlation effects} \end{array}\right)$$

$$+ \left(\frac{\partial r}{\partial x}\right)^2 s_x^2 + \left(\frac{\partial r}{\partial y}\right)^2 s_y^2 + \left(\begin{array}{c} \text{random error} \\ \text{correlation effects} \end{array}\right) \tag{3.6}$$

where the b_x and b_y systematic standard uncertainties are determined from the combination of elemental systematic uncertainties that affect x and y as

$$b_x = \sum_{k=1}^{M_x} b_{x_k}^2 \tag{3.7}$$

and

$$b_y = \sum_{k=1}^{M_y} b_{y_k}^2 \qquad (3.8)$$

The summation parameters M_x and M_y are the number of significant elemental systematic error sources for x and y, respectively. The random standard uncertainties s_x and s_y are the standard deviations for the measurements of x and y.

Correlation terms are included in Eq. (3.6) because in some cases errors may not be independent. If all of the systematic error sources for x and y were totally independent of each other, then the systematic error correlation effects term would be zero. But in some cases, x and y may share a common error source, as in a calibration standard error that is the same for both. For instance, x and y could be temperature measurements made by separate thermocouples. If the two thermocouples were calibrated against the same standard, then the two temperature measurements would have a common, identical, systematic error resulting from the accuracy of the calibration standard. The systematic error correlation term in Eq. (3.6) allows for the correction of the combined standard uncertainty to take into account the effects of these correlated errors. This topic will be discussed in detail in Chapter 5.

Normally, we consider random errors to be uncorrelated and would ignore the random error correlation effects term in Eq. (3.6). However, as will be shown in Chapter 5, there often can be effects that vary with time that affect x and y in the same manner. For instance, consider a result that is the pressure drop for flow in a pipe determined by taking the difference in the inlet and exit pressures. For the steady-state calculation of the pressure drop, s_x and s_y are the standard deviations of the inlet and exit pressure measurements. If there were some external effect causing the fluid flow rate to vary during the measurements, then the inlet and exit pressures would be affected by the varying flow rate in the same way. They would have a common, time-varying error source. The methods for handling correlated, time-varying errors are given in Chapter 5. For the remainder of this section, we will leave the correlation terms, both systematic and random, out of the TSM equations and then consider them in detail in Chapter 5.

The TSM expression in Eq. (3.6) is an approximation that includes only the first-order terms from the Taylor series expansion. Also, the derivatives in Eq. (3.6) are evaluated at the measured values of x and y instead of at x_{true} and y_{true}, which would be required for a formal Taylor series expansion. For most engineering and scientific applications, these approximations are reasonable, and the resulting combined standard uncertainty u_r is a good estimate of the standard deviation for the parent population for the result, r.

For the more general case where the result is a function of several variables

$$r = r(X_1, X_2, \ldots, X_J) \qquad (3.9)$$

the combined standard uncertainty is given by

$$u_r^2 = \sum_{i=1}^{J} \left(\frac{\partial r}{\partial X_i} \right)^2 b_{X_i}^2 + \sum_{i=1}^{J} \left(\frac{\partial r}{\partial X_i} \right)^2 s_{X_i}^2 \qquad (3.10)$$

Each b_{X_i} is determined as shown in Eqs. (3.7) and (3.8), and the s_{X_i} values are the standard deviations for the measurement of each variable, X_i.

The first group of terms on the right-hand side of Eq. (3.10) can be defined as the systematic standard uncertainty of the result, b_r, as

$$b_r^2 = \sum_{i=1}^{J} \left(\frac{\partial r}{\partial X_i} \right)^2 b_{X_i}^2 \qquad (3.11)$$

This definition will be useful as we move into the design phase of an experiment, as presented in Chapter 5. Likewise, the random standard uncertainty of the result, s_r, can be defined from Eq. (3.10) as

$$s_r^2 = \sum_{i=1}^{J} \left(\frac{\partial r}{\partial X_i} \right)^2 s_{X_i}^2 \qquad (3.12)$$

The expressions in Eqs. (3.10) through (3.12) would have to include the correlation terms if they were applicable.

Later, in the debugging and execution phases of the experimental program, we will find that once we have data to calculate the result from Eq. (3.9), M multiple results at a given test condition can be used to calculate the random standard uncertainty of the result directly as

$$s_r^2 = \frac{1}{M-1} \sum_{k=1}^{M} (r_k - \bar{r})^2 \qquad (3.13)$$

where

$$\bar{r} = \frac{1}{M} \sum_{k=1}^{M} r_k \qquad (3.14)$$

We will find that if some of the X_i variables have a common, time-varying error source, then the random error correlation effects will automatically be taken into account when we use Eq. (3.13) for s_r. Thus it is preferable to directly calculate s_r using Eq. (3.13) and perhaps compare it with the value calculated using Eq. (3.12) when sufficient information is available.

Using the defined systematic and random standard uncertainties of the result, the combined standard uncertainty in Eq. (3.10) becomes

$$u_r = \left(b_r^2 + s_r^2 \right)^{1/2} \qquad (3.15)$$

Next we discuss how to expand this standard deviation level estimate of the uncertainty in the result to a specified percent level of confidence.

3-1.2 Expanded Uncertainty of a Result

In Section 2-5.2, we presented the methodology for determining the overall or expanded uncertainty of a measured variable. The approach for a result which is a function of several variables is the same and is based on the central limit theorem.

As recommended in the ISO guide [1], the expanded uncertainty at a given percent level of confidence can be determined using a coverage factor such that

$$U_r = k_\% u_r \tag{3.16}$$

The central limit theorem tells us that the error distribution for r will usually approach Gaussian, allowing us to use the t distribution (Table A.2) to obtain $k_\%$. Therefore, the expanded uncertainty for a given percent level of confidence will be

$$U_r = t_\% u_r \tag{3.17}$$

where the $\pm U_r$ band around r will contain the true value of the result at that level of confidence. Substituting Eq. (3.15) for u_r, the expanded uncertainty becomes

$$U_r = t_\% \left(b_r^2 + s_r^2 \right)^{1/2} \tag{3.18}$$

The value $t_\%$ from the t distribution requires a degrees of freedom for the result. As with the single variable [Eq. (2.28)], the degrees of freedom of the result, ν_r, is dependent on the degrees of freedom of all the elemental standard uncertainty sources. In the case of a result, we have all the elemental systematic standard uncertainty sources for each variable in addition to the random standard uncertainties. The Welch–Satterthwaite approximation for the degrees of freedom of the result is given in Appendix B [Eq. (B.19)], with the more general form given in Appendix C [Eq. (C.4)]. As we saw for the single variable in Chapter 2, the degrees of freedom for most engineering and scientific results will usually be large enough to consider t to be a constant for a given level of confidence. This large-sample approximation is discussed next.

3-1.3 Large-Sample Approximation for Uncertainty of a Result

The authors conducted a study to investigate the effects of the degrees of freedom of the result on the expansion factor $t_\%$ in Eq. (3.18). In the study we looked at a range of DREs from simple linear expressions to those that were highly nonlinear. The methodology and conclusions of the study are given in Appendix C. We assumed that we knew the true values for each of the variables and, therefore, the true value of the result. Monte Carlo simulations were used to test various expansion factor assumptions to see which ones would provide an uncertainty interval that contained the true result with a given level of confidence.

In the study, various cases were run taking the systematic and random uncertainty effects to be balanced, considering one or the other to be dominant, or assuming that the uncertainty for one of the variables in the DRE dominated the uncertainty of the result. In all of these examples, cases were run with the degrees of freedom for the random and systematic uncertainties varying from 2 to 31. For each case, the degrees of freedom of the result was calculated using the Welch–Satterthwaite approximation in Eq. (C.4). Then the actual percent coverage for the true result was determined using either t_{95} or t_{99} for the appropriate degrees of freedom of the result as the expansion factor or using the large-sample constant values of 2 or 2.6.

The results of all of the simulations are shown in Figure 3.1 for the uncertainty models that were intended to yield a 95% level of confidence (t_{95} or $t = 2$) for the cases where the degrees of freedom of the result, ν_r, was approximately 9. We see that even when we use t_{95} for the appropriate degrees of freedom as the expansion factor, we do not necessarily get 95% coverage of the true result. The t_{95} cases showed a peak at about 96% coverage and the $t = 2$ cases peaked at about 93.5% coverage. The point is that neither model gave exactly 95% coverage, but both gave coverages in the mid-90s even for 9 degrees of freedom in the result.

When the degrees of freedom for each random standard uncertainty and each elemental systematic standard uncertainty is 9, the simulation cases overlap, as shown in Figure 3.2. Both models gave coverages that peaked near 95%. Of course, when the degrees of freedom of all of the uncertainty components are 9, the degrees of freedom of the result will be much greater from Eq. (C.4).

Similar conclusions for a 99% level of confidence were found as shown in Figures C.1 and C.3. For 9 degrees of freedom in the result, t_{99} and $t = 2.6$ both

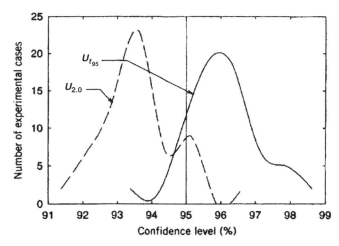

Figure 3.1 Histogram of levels of confidence provided by 95% uncertainty models for all experiments where $\nu_r \approx 9$ (Appendix C).

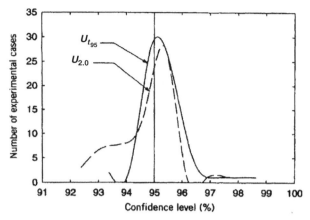

Figure 3.2 Histogram of levels of confidence provided by 95% uncertainty models for all experiments when $\nu_{s_i} = \nu_{b_{i_k}} = 9$ (Appendix C).

gave coverages in the high-90s. For 9 degrees of freedom for all uncertainties, both peaked at about 99% for the cases run.

The conclusion from the study in Appendix C is that for most engineering and scientific applications the large-sample assumption will be appropriate so that $t = 2$ for a 95% level of confidence and $t = 2.6$ for a 99% level of confidence. Therefore, from Eq. (3.18), the large-sample expression for the expanded uncertainty of the result at a 95% level of confidence is given as

$$U_{95} = 2\left(b_r^2 + s_r^2\right)^{1/2} \tag{3.19}$$

Since many uncertainty applications use a 95% level of confidence, we will use Eq. (3.19) for the experiment design, debugging, and execution phase discussions in Chapter 5 and for the regression uncertainty presentation in Chapter 7. We call the use of Eq. (3.19) *detailed uncertainty analysis*.

If we substitute b_r and s_r from Eqs. (3.11) and (3.12) into Eq. (3.19), the large-sample 95% expanded uncertainty is

$$U_{95} = 2\left[\sum_{i=1}^{J}\left(\frac{\partial r}{\partial X_i}\right)^2\left(b_{X_i}^2 + s_{X_i}^2\right)\right]^{1/2} \tag{3.20}$$

If we assume that there are no correlated systematic or random errors, then all of the b_{X_i} and s_{X_i} terms are independent. Taking the coverage factor inside the summation, we get

$$U_{95} = \left[\sum_{i=1}^{J}\left(\frac{\partial r}{\partial X_i}\right)^2 2^2 u_i^2\right]^{1/2} \tag{3.21}$$

or

$$U_{95} = \left[\sum_{i=1}^{J} \left(\frac{\partial r}{\partial X_i} \right)^2 U_i^2 \right]^{1/2} \tag{3.22}$$

where each U_i is the large-sample 95% expanded uncertainty for variable X_i. This equation describes the propagation of the overall uncertainties in the measured variables into the overall uncertainty of the result. The application of this equation is termed *general uncertainty analysis* and is used in Chapter 4 in the discussion of the planning phase of an experiment or simulation.

3-1.4 Example of TSM Uncertainty Propagation

Consider the example given in Section 1-4 for the determination of the resistance characteristics of a rough-walled pipe in a flow system. The schematic of the system is shown in Figure 3.3 along with nominal values of the variables for a specific experimental run. The pipe resistance is expressed as a friction factor f,

$$f = \frac{\pi^2 d^5 \Delta P}{8 \rho Q^2 L} \tag{3.23}$$

for a given flow condition represented by the Reynolds number

$$\text{Re} = \frac{4Q}{\pi \mu D} \tag{3.24}$$

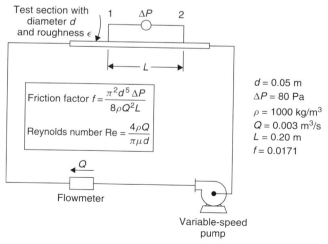

Figure 3.3 Experimental determination of resistance characteristics of rough-walled pipe.

The variables are pipe diameter d, pressure drop ΔP over a length L, fluid volumetric flow rate Q, fluid density ρ, and dynamic viscosity μ.

Using the general uncertainty analysis expression (3.22) for the friction factor, the uncertainty is determined as

$$U_f^2 = \left(\frac{\partial f}{\partial d}\right)^2 U_d^2 + \left(\frac{\partial f}{\partial \Delta P}\right)^2 U_{\Delta P}^2 + \left(\frac{\partial f}{\partial \rho}\right)^2 U_\rho^2$$
$$+ \left(\frac{\partial f}{\partial Q}\right)^2 U_Q^2 + \left(\frac{\partial f}{\partial L}\right)^2 U_L^2 \tag{3.25}$$

Taking the derivatives of f with respect to the variables, substituting into Eq. (3.25), and then dividing by f^2, we get

$$\left(\frac{U_f}{f}\right)^2 = (5)^2 \left(\frac{U_d}{d}\right)^2 + (1)^2 \left(\frac{U_{\Delta P}}{\Delta P}\right)^2 + (1)^2 \left(\frac{U_\rho}{\rho}\right)^2 + (2)^2 \left(\frac{U_Q}{Q}\right)^2$$
$$+ (1)^2 \left(\frac{U_L}{L}\right)^2 \tag{3.26}$$

If the uncertainty for each variable is taken as 1% at a 95% level of confidence

$$\frac{U_{X_i}}{X_i} = 0.01 \tag{3.27}$$

then the uncertainty for the friction factor using the TSM is

$$\left(\frac{U_f}{f}\right)_{\text{TSM}} = 5.7\% \tag{3.28}$$

or

$$(U_f)_{\text{TSM}} = 0.00097 \tag{3.29}$$

where

$$f = 0.0171 \tag{3.30}$$

from Eq. (3.23) using the nominal input values. For 5% uncertainty for each variable, the uncertainty in f is

$$\left(\frac{U_f}{f}\right)_{\text{TSM}} = 28.3\% \tag{3.31}$$

or

$$(U_f)_{\text{TSM}} = 0.00484 \tag{3.32}$$

3-1.5 Numerical Approximation for TSM Propagation of Uncertainties

Recall from Section 3-1.3 that, if an experimental result is given by the data reduction equation (3.9),

$$r = r(X_1, X_2, \ldots, X_J)$$

the 95% expanded uncertainty in the result for the general uncertainty analysis method can be expressed as Eq. (3.22):

$$U_r = \left[\left(\frac{\partial r}{\partial X_1} U_{X_1} \right)^2 + \left(\frac{\partial r}{\partial X_2} U_{X_2} \right)^2 + \cdots + \left(\frac{\partial r}{\partial X_J} U_{X_J} \right)^2 \right]^{1/2}$$

There are cases when the data reduction expression is very complex and the task of obtaining the partial derivatives for the TSM propagation equation is extremely laborious. When a case is encountered such that the chances of obtaining correct analytical expressions for the partial derivatives is remote or, indeed, the expression (3.9) is so intimidating that one might be tempted to proceed without performing an uncertainty analysis, numerical approximations to the partial derivatives should be used. Using a forward-differencing finite-difference approach as an example, we can write

$$\left. \frac{\partial r}{\partial X_1} \right|_{X_2, \ldots, X_J \text{const}} \approx \frac{\Delta r}{\Delta X_1} \approx \frac{r_{X_1 + \Delta X_1, X_2, \ldots, X_J} - r_{X_1, X_2, \ldots, X_J}}{\Delta X_1} \tag{3.33}$$

with similar expressions for the derivatives with respect to X_2 and X_J. An initial estimate of ΔX_i, might be $0.01X_i$. The numerical derivative is calculated with this first perturbation and then with a perturbation of one-half that first perturbation. These two derivative values are compared for convergence, and the process is repeated if necessary. Once converged derivative estimates are obtained, the finite-difference approximation to Eq. (3.22) then becomes

$$U_r^2 \approx \left(\frac{\Delta r}{\Delta X_1} U_{X_1} \right)^2 + \left(\frac{\Delta r}{\Delta X_2} U_{X_2} \right)^2 + \cdots + \left(\frac{\Delta r}{\Delta X_J} U_{X_J} \right)^2 \tag{3.34}$$

This type of approach is fairly easily implemented using a spreadsheet or computer program [2]. Also, commercial mathematics packages such as Mathead use a similar approach to obtain values for derivatives [3].

Other methods for computing the derivatives which converge faster include the central-difference approach,

$$\left. \frac{\partial r}{\partial X_1} \right|_{X_2, \ldots, X_J \text{const}} = \frac{r_{X_1 + \Delta X_1, X_2, \ldots, X_J} - r_{X_1 - \Delta X_1, X_2, \ldots, X_J}}{2 \, \Delta X_1} \tag{3.35}$$

If complex variables can be used for the X_i's, then

$$\left. \frac{\partial r}{\partial X_1} \right|_{X_2, \ldots, X_J \text{const}} = \frac{\text{Im} \left(r_{X_1 + \Delta X_1, X_2, \ldots, X_J} \right)}{\Delta X_1} \tag{3.36}$$

where Im $\left(r_{X_1+\Delta X_1, X_2, \ldots, X_J}\right)$ is the imaginary part of the result, r [4]. The advantage of this complex-step method is that very small values of ΔX_1 can be used without affecting the accuracy of the approximation for the derivative. The iterations required in Eqs. (3.33) and (3.35) are not necessary. This is very important in simulation uncertainty determinations, in computational fluid dynamics (CFD) for example, where one solution may take many hours of computer time.

3-2 MONTE CARLO METHOD FOR PROPAGATION OF UNCERTAINTIES

An introduction to the Monte Carlo Method (MCM) was given in Section 2-5.4 for a measured variable. That discussion illustrated the random sampling from assumed distributions for each error source to estimate the distribution of the measured variable. With the computing power and speed now available, even in personal computers, it has become feasible to perform an uncertainty analysis directly by Monte Carlo simulation for a result that is a function of multiple variables. This method is not limited to simple expressions but can also be used for highly complicated experiment data reduction equations or for numerical solutions of advanced simulation equations. The Joint Committee for Guides in Metrology (JCGM) has recently published a supplement to the GUM [1] presenting the MCM for uncertainty analysis [5].

In this section, we will present the general approach for the MCM for uncertainty propagation and will then illustrate the technique with an example. The section concludes with a discussion of the general methodology for determining a specific percent coverage interval for the true result from the Monte Carlo distribution of results.

3-2.1 General Approach for MCM

Figure 3.4 presents a flowchart that shows the steps involved in performing an uncertainty analysis by the MCM. The figure shows the sampling technique for a function of two variables, but the methodology is general for DREs or simulations that are functions of multiple variables. In many cases, more than one result is desired in the test or simulation, that is, pressure drop and friction factor for a given Reynolds number, and that possibility is illustrated in the figure.

Figure 3.4 illustrates the MCM case that is analogous to the TSM case in which random standard uncertainties s_{X_i} are used in the propagation equation (3.12) to obtain s_r. The MCM case analogous to the TSM case in which s_r is directly calculated from multiple results using Eq. (3.13) is shown in Figure 3.5.

Using the approach shown in Figure 3.4 for MCM propagation as the basis for the discussion, first the assumed true value of each variable in the result is input. These would be the X_{best} values that we have for each variable. Then the estimates of the random standard uncertainty s and the elemental systematic standard uncertainties b_k for each variable are input. An appropriate probability distribution function is assumed for each error source. The random errors are usually assumed to come from a Gaussian distribution, but if s is from a very

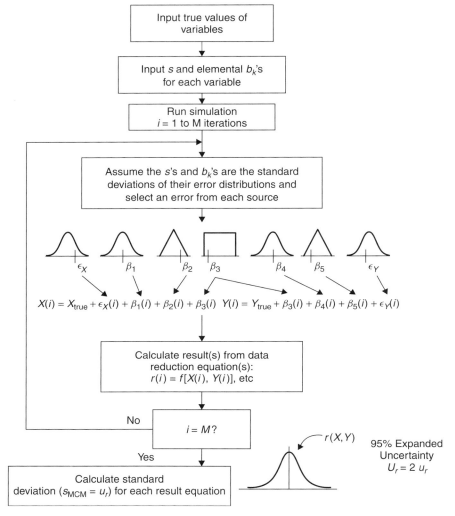

Figure 3.4   Schematic of MCM for uncertainty propagation when random standard uncertainties for individual variables are used.

small sample, the t distribution at the appropriate degrees of freedom ν_s could be more appropriate. The systematic error distributions are chosen based on the user's judgment, as discussed in Section 2-5.1. For the flowchart in Figure 3.4, we have assumed that the random standard uncertainties for X and Y come from Gaussian distributions and that each variable has three elemental systematic standard uncertainties, one Gaussian, one triangular, and one rectangular.

For each variable, random values for the random errors and each elemental systematic error are found using an appropriate random number generator (Gaussian,

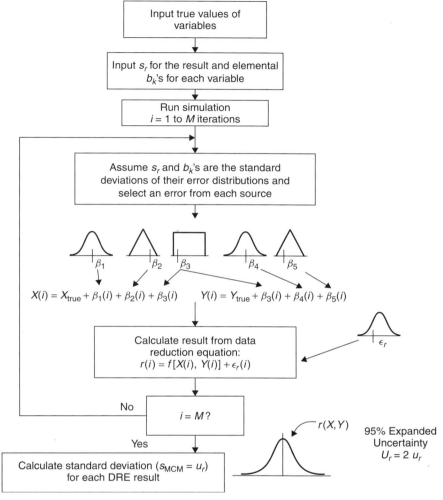

Figure 3.5   Schematic of MCM for uncertainty propagation when directly calculated random standard uncertainty for the result, s_r, is used.

rectangular, triangular, etc.). The individual error values are then summed and added to the true values of the variables to obtain "measured" values. Using these measured values, the result is calculated. This process corresponds to running the test or simulation once.

The sampling process is repeated M times to obtain a distribution for the possible result values. The primary goal of the MCM propagation technique is to estimate a converged value for the standard deviation, s_{MCM}, of this distribution. An appropriate value for M is determined by periodically calculating s_{MCM} during the MCM process and stopping the process when a converged value of s_{MCM} is obtained. Keep in mind that s_{MCM} is the estimate of the combined standard uncertainty of the result, u_r. We do not need to have a perfectly converged value of s_{MCM} to have a reasonable estimate of u_r. Once the s_{MCM} values are converged to within 1–5%, then the value of s_{MCM} is a good approximation of the combined standard uncertainty of the result. The level of convergence is a matter of judgment based on the cost of the sampling process (e.g., costly CFD simulations) and the application for u_r. Once a converged value of u_r is determined and assuming that the central limit theorem applies, the expanded uncertainty for the result at a 95% level of confidence is $U_r = 2u_r$.

The number of iterations required in the MCM to reach a reasonable estimate of the converged value of s_{MCM} can be reduced significantly by using special sampling techniques, such as Latin hypercube sampling. This technique is described in Ref. 6–10 and is very popular for determining the uncertainty in complex simulation results.

Note that in Figure 3.4 we used the same value of β_3 for X and Y during each iteration. Here we are assuming that X and Y have a common, correlated, systematic error that has the same value for both variables. As we discussed in Section 3-1.1, the effects of these correlated errors have to be taken into account in order to get the correct value for u_r. We see in Figure 3.4 that handling correlated systematic errors is very straightforward using the MCM.

In Section 3-1.1, we also pointed out that sometimes the random errors might be correlated because of effects that vary with time that affect the variables in the same manner. These effects are taken into account using the approach shown in Figure 3.5. In the next section, we give an example of using the MCM for uncertainty propagation.

3-2.2 Example of MCM Uncertainty Propagation

Considering the same example used in Section 3-1.4, the uncertainties of the three test results, friction factor f, Reynolds number Re, and relative roughness ϵ/d, can be found with the MCM method as shown in Figure 3.6. The nominal values for all of the variables are input to the MCM process. Standard uncertainties and associated error distributions are assumed for each variable. Error values are randomly chosen from the distributions and added to the nominal values to get the measurement values for each variable and then the values of the three results. The iteration process is continued until converged values of u_f, u_{Re}, and $u_{e/d}$ are obtained. In this example, we have taken one error source for each variable to simplify the presentation. Of course, the MCM process is straightforward for multiple error sources for each variable, as shown in the previous section.

In the TSM solution of this problem in Section 3-1.4, percent expanded uncertainties at a 95% level of confidence were used for each of the variables. If we assume that all distributions are Gaussian, then the standard uncertainty inputs for the MCM will be the previous percent uncertainties divided by 2. Therefore for 1% expanded uncertainties

$$u_{X_i} = \tfrac{1}{2}(0.01)X_i \tag{3.37}$$

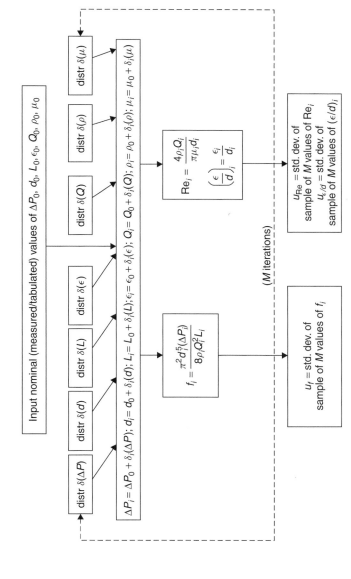

Figure 3.6 Schematic flowchart of MCM for fluid flow example problem.

75

Using the same nominal values for the variables as before, the MCM yields for the friction factor

$$(u_f)_{\text{MCM}} = 0.00049 \tag{3.38}$$

or

$$\left(\frac{U_f}{f}\right)_{\text{MCM}} = \left(\frac{2u_f}{f}\right)_{\text{MCM}} = 5.7\% \tag{3.39}$$

which is the same value as that obtained using the TSM. The distribution of the results from the MCM is shown in Figure 3.7.

The convergence of the combined standard uncertainty u_f in this MCM example is shown in Figure 3.8. The value of s_f was calculated after 200 iterations and then again for every 200 additional iterations using the total number of iterations to that point. The plot of these s_f, or u_f, values is shown in Figure 3.8. The value of $(U_f/f)_{\text{MCM}}$ in Eq. (3.39) was a fully converged value from a large number of iterations. We see that after only 200 iterations the value has converged to within about 3% and by 1000 iterations to within less than 1% of the fully converged value.

When the 95% expanded uncertainties for each variable are 5%, the MCM results for the friction factor yield

$$(u_f)_{\text{MCM}} = 0.00244 \tag{3.40}$$

or

$$\left(\frac{U_f}{f}\right)_{\text{MCM}} = \left(\frac{2u_f}{f}\right)_{\text{MCM}} = 28.5\% \tag{3.41}$$

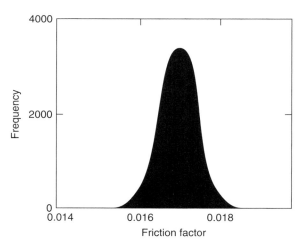

Figure 3.7 Distribution of MCM results for friction factor for 1% expanded uncertainty (95%) for each variable.

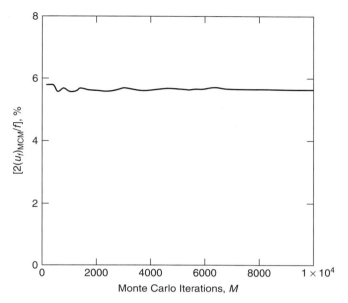

Figure 3.8 Convergence study for MCM value of u_f for 1% expanded uncertainty (95%) for each variable.

This value is compared to the value for $(U_f/f)_{TSM}$ of 28.3%, which is essentially the same value. Looking at the distribution of the results for this case from the MCM given in Figure 3.9, we see that when the expanded uncertainties of the variables are 5% the distribution of MCM results is slightly skewed toward higher values of friction factor.

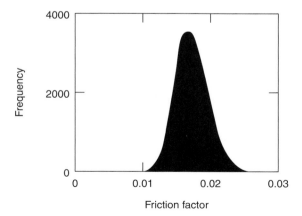

Figure 3.9 Distribution of MCM results for friction factor for 5% expanded uncertainties (95%) for each variable.

The difference between the MCM and TSM uncertainty results in this example is negligible, but if the distribution of MCM results is highly skewed, the plus and minus uncertainty limits will not be equal. The general methodology for determining a specific percent coverage interval for the true result from the MCM is presented next.

3-2.3 Coverage Intervals for MCM Simulations

If the uncertainties of the variables are relatively large and/or the DRE or simulation is highly nonlinear, the distribution of Monte Carlo results can be asymmetric, as shown in Figure 3.10. The result calculated using the nominal values of all the input variables will not coincide with the peak, or mode (the most probable value), of the distribution of the MCM results as shown. For these cases, calculating the standard deviation s_{MCM} and assuming that the central limit theorem applies to obtain the expanded uncertainty will not necessarily be appropriate depending on the degree of asymmetry.

The GUM supplement [5] presents methods for determining the coverage intervals for MCM simulations when the distributions are asymmetric. These coverage intervals are described below.

The coverage interval that provides "probabilistically symmetric coverage" is shown in Figure 3.11 for a 95% level of confidence. The interval contains $100p\%$ of the results of the MCM simulation, where in this case $p = 0.95$ for 95% coverage. The lower bound of the interval is the result value that corresponds to the upper end of the lowest 2.5% of the MCM results. The upper bound of the interval is the result that corresponds to the upper end of the range that contains 97.5% of the MCM results. The interval between these bounds contains 95% of the MCM results with a symmetric percentage of the MCM results (2.5% in this case) outside of the coverage interval on each end.

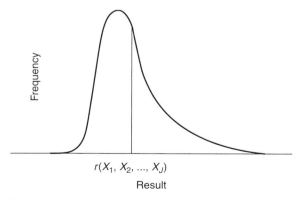

Figure 3.10 MCM results for large uncertainties in variables and/or nonlinear result equations.

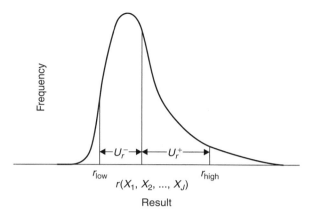

Figure 3.11   Probabilistically symmetric coverage interval for 95% level of confidence.

The procedure for determining the 95% probabilistically symmetric coverage interval and associated uncertainty limits is given below:

1. Sort the M Monte Carlo simulation results from lowest value to the highest value.
2. For a 95% coverage interval

$$r_{\text{low}} = \text{result number}(0.025M) \tag{3.42}$$

and

$$r_{\text{high}} = \text{result number}(0.975M) \tag{3.43}$$

   If the numbers $0.025M$ and $0.975M$ are not integers, then add $\frac{1}{2}$ and take the integer part as the result number [5].
3. For 95% expanded uncertainty limits

$$U_r^- = r(X_1, X_2, \ldots, X_J) - r_{\text{low}} \tag{3.44}$$

and

$$U_r^+ = r_{\text{high}} - r(X_1, X_2, \ldots, X_J) \tag{3.45}$$

4. The interval that contains r_{true} at a 95% level of confidence is then

$$r - U_r^- \leq r_{\text{true}} \leq r + U_r^+ \tag{3.46}$$

The general procedure for determining the probabilistically symmetric coverage interval for any $100p\%$ level of confidence ($p = 0.99$ for 99% for instance) is step 1 above followed by

$$r_{\text{low}} = \text{result number}\{[\tfrac{1}{2}(1 - p)]M\} \tag{3.47}$$

and

$$r_{\text{high}} = \text{result number}\{[\tfrac{1}{2}(1+p)]M\} \qquad (3.48)$$

Once again, if these numbers are not integers, then $\tfrac{1}{2}$ is added to each and the integer part is used as the result number. The expressions for the asymmetric uncertainty limits, U_r^- and U_r^+, and the $100p\%$ coverage interval for r_{true} are the same as in steps 3 and 4 above.

The GUM supplement [5] provides an alternative coverage interval called the "shortest coverage interval," which is described in Appendix D. If the distribution of MCM results is symmetric, the shortest coverage interval, the probabilistically symmetric coverage interval, and the interval given by $\pm(k_{100p\%})s_{\text{MCM}}$ will be the same for a $100p\%$ level of confidence.

3-2.4 Example of Determination of MCM Coverage Interval

Consider again the example in Sections 3-1.4 and 3-2.2. We saw in Section 3-2.2 that when the 95% expanded uncertainties for each variable were 5% the 95% expanded uncertainty for the friction factor was 28.5% using $2s_{\text{MCM}}/f$. The uncertainty value from the TSM was $(U_f/f)_{\text{TSM}} = 28.3\%$. We also observed that the distribution of friction factor results for this case was slightly skewed toward larger values, as shown in Figure 3.9.

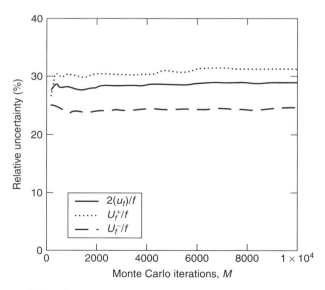

Figure 3.12 Convergence of coverage intervals for MCM example.

Using the probabilistically symmetric coverage interval method, $p = 0.95$, and the distribution of friction factor results in Figure 3.9, the plus and minus uncertainty limits were obtained as

$$U_f^+ = 31.5\% \tag{3.49}$$

and

$$U_f^- = 24.5\% \tag{3.50}$$

These values along with the value in Eq. (3.41) represent the converged uncertainty limits for a large number of iterations. The convergence of these limits is shown in Figure 3.12. We see that with about 2000 iterations the values for U_f^- and U_f^+ have converged to within about 5% and by 10,000 iterations to within about 2% of the fully converged values.

For this example, the TSM uncertainty interval, the $2(u_f)_{\mathrm{MCM}}/f$ interval, and the asymmetric uncertainty interval all provide about 95% coverage of the results from the MCM distribution given in Figure 3.9. However, for extreme cases where the Monte Carlo distribution is highly skewed, the asymmetric uncertainty limits will be more appropriate to provide a given level of confidence for the uncertainty estimate.

REFERENCES

1. International Organization for Standardization (ISO), *Guide to the Expression of Uncertainty in Measurement*, ISO, Geneva, 1993.
2. Steele, W. G., Ferguson, R. A., Taylor, R. P., and Coleman, H. W., "Computer-Assisted Uncertainty Analysis," *Computer Applications in Engineering Education*, Vol. 5, No. 3, 1997, pp. 169–179.
3. Hosni, M. H., Coleman, H. W., and Steele, W. G., "Application of MathCad Software in Performing Uncertainty Analysis Calculations to Facilitate Laboratory Instruction," *Computers in Education Journal*, Vol. 7, No. 4, Oct.–Dec. 1997, pp. 1–9.
4. Newman III, J. C., Whitfield, D. L., and Anderson, W. K., "Step-Size Independent Approach for Multidisciplinary Sensitivity Analysis," *Journal of Aircraft*, Vol. 40, No. 3, May–June, 2003, pp. 566–573.
5. Joint Committee for Guides in Metrology (JCGM), "Evaluation of Measurement Data—Supplement 1 to the 'Guide to the Expression of Uncertainty in Measurement'—Propagation of Distributions Using a Monte Carlo Method," JCGM 101:2008, France, 2008.
6. McKay, M. D., Conover, W. J., and Beckman, R. J., "A Comparison of Three Methods for Selecting Values of Input Variables in the Analysis of Output from a Computer Code," *Technometrics*, Vol. 21, 1979, pp. 239–245.
7. Helton, J. C., and Davis, F. J., "Latin Hypercube Sampling and the Propagation of Uncertainty in Analyses of Complex Systems," *Reliability Engineering and System Safety*, Vol. 81, 2003, pp. 23–69.

8. Helton, J. C., and Davis, F. J., "Sampling Based Methods," in Saltelli, A., Chan, K., and Scott, E. M., Eds., *Sensitivity Analysis*, Wiley, New York, 2000, pp. 101–153.

9. Iman, R. L., and Shortencarier, M. J., "A FORTRAN 77 Program and User's Guide for the Generation of Latin Hypercube and Random Samples for Use with Computer Models," Technical Report SAND83-2356, Sandia National Laboratories, Albuquerque, NM, Mar. 1984.

10. Wyss, G. D., and Jorgensen, K. H., "A User's Guide to LHS: Sandia's Latin Hypercube Sampling Software," Technical Report SAND98-0210, Sandia National Laboratories, Albuquerque, NM, Feb. 1998.

PROBLEMS

3.1 Below is a sketch of a turbine showing the measurements of the inlet and exit pressures and temperatures:

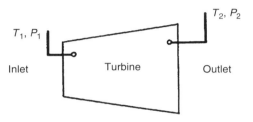

The expression for the adiabatic turbine efficiency η is

$$\eta = \frac{1 - (T_2/T_1)}{1 - (P_2/P_1)^{(\gamma-1)/\gamma}}$$

where γ is the ratio of specific heats. Use the following nominal values and total uncertainty estimates for the variables:

Variable	Nominal Value	Uncertainty (95%)
P_1	517 kPa	10%
P_2	100 kPa	10%
T_1	1273 K	3%
T_2	853 K	3%
γ	1.4	0.5%

Perform a general uncertainty analysis using TSM propagation to estimate the uncertainty in the adiabatic turbine efficiency.

3.2 Perform the analysis in problem 3.1 using MCM propagation and compare the two uncertainty estimates.

3.3 Repeat problems 3.1 and 3.2 with the uncertainties taken as 3 times the given values. Compare the TSM and MCM results using $2u_\eta$ for the MCM uncertainty at a 95% level of confidence.

3.4 Use the methods of Section 3-2.3 to determine the MCM asymmetric uncertainty limits for problem 3.3 and compare the results with the $2u_\eta$ value.

4

GENERAL UNCERTAINTY ANALYSIS: PLANNING AN EXPERIMENT AND APPLICATION IN VALIDATION

4-1 OVERVIEW: USING UNCERTAINTY PROPAGATION IN EXPERIMENTS AND VALIDATION

As we discussed in Chapter 3, when a result is determined from the values of other variables as in

$$r = r(X_1, X_2, \ldots, X_J) \qquad (4.1)$$

we must consider how the uncertainties in the X_i's propagate into the result r. If we are considering an experiment, then Eq. (4.1) is a data reduction equation and the X_i's are the measured variables or values found from reference sources (in the case of material properties not measured in the experiment). If we are considering a simulation, then Eq. (4.1) symbolically represents the model that is being solved and the X_i's are the model inputs (geometry, properties, etc.)—(4.1) might be a single expression or it might be thousands of lines of computer code.

In Chapter 3, two propagation methods—the TSM and the MCM—were introduced as were the ideas of *detailed uncertainty analysis* and *general uncertainty analysis*. We consider the application of general uncertainty analysis in this chapter and the application of detailed uncertainty analysis in Chapter 5. In the following, Sections 4-2 through 4-7 consider the use of the TSM in the application of general uncertainty analysis in experiments. In Section 4-8, the application of the general uncertainty equations is considered when Eq. (4.1) represents a simulation.

4-1.1 Application in Experimentation

What uncertainty should be associated with experimental results determined using a data reduction equation? The measurements of the variables (temperature, pressure, velocity, etc.) have uncertainties associated with them, and the values of the material properties that we obtain from reference sources also have uncertainties. *How do the uncertainties in the individual variables propagate through a data reduction equation into a result?* This is a key question in experimentation, and its answer is found using uncertainty analysis.

In the planning phase of an experimental program, the approach we use considers only the general, or overall, uncertainties and not the details of the systematic and random components. This approach is termed *general uncertainty analysis*. It makes sense to consider only the overall uncertainty in each measured variable at this stage rather than worry about which part of the uncertainty will be due to systematic and which part will be due to random effects. In what we are calling the planning phase, we are trying to identify which experiment or experiments will give us a chance to answer the question we have. In general, the particular equipment and instruments will not have been chosen—whether a temperature, for instance, will be measured with a thermocouple, thermometer, thermistor, or optical pyrometer is yet to be decided.

In a sense, at this stage all contributors to the uncertainty can be considered to be random. Since no particular transducer or instrument has been chosen from the population of all transducers and instruments available, any systematic error is just as likely to be positive as negative. Thus we can view the general uncertainty analysis performed in the planning phase as a special case in which, for each measured variable X_i, $b_{X_i} = 0$ and therefore the uncertainty is random. Of course, at this stage there are usually no samples from which to compute statistical estimates of s_{X_i} using the methods of Chapter 2. Often, parametric studies are made for an assumed range of uncertainties in the variables. This is demonstrated later in this chapter.

In the following sections we describe the technique of general uncertainty analysis and some helpful hints for its application, discuss the factors that must be considered in applying uncertainty analysis in the planning phase of an experimental program, and present comprehensive examples of the application of general uncertainty analysis in planning an experiment.

4-2 GENERAL UNCERTAINTY ANALYSIS USING THE TAYLOR SERIES METHOD (TSM)

Consider a general case in which an experimental result r is a function of J measured variables X_i:

$$r = r(X_1, X_2, \ldots, X_J) \tag{4.2}$$

Equation (4.2) is the data reduction equation used for determining r from the measured values of the variables X_i. Then the uncertainty in the result is given by

$$U_r^2 = \left(\frac{\partial r}{\partial X_1}\right)^2 U_{X_1}^2 + \left(\frac{\partial r}{\partial X_2}\right)^2 U_{X_2}^2 + \cdots + \left(\frac{\partial r}{\partial X_J}\right)^2 U_{X_J}^2 \qquad (4.3)$$

where the U_{X_i} are the uncertainties in the measured variables X_i.

The development of Eq. (4.3) is presented and discussed in Appendix B. It is assumed that the relationship given by (4.2) is continuous and has continuous derivatives in the domain of interest, that the measured variables X_i are independent of one another, and that the errors in the measured variables are independent of one another.

The partial derivatives are called *absolute sensitivity coefficients*, and sometimes nomenclature is used such that

$$\theta_i = \frac{\partial r}{\partial X_i} \qquad (4.4)$$

Eq. (4.3) can be written as

$$U_r^2 = \sum_{i=1}^{J} \theta_i^2 U_{X_i}^2 \qquad (4.5)$$

In Eqs. (4.3) and (4.5), all of the *absolute uncertainties* (U_{X_i}) should be expressed with the same odds or level of confidence. In most cases, 95% confidence (20:1 odds) is used, with the uncertainty in the result then also being at 95% confidence. In the planning phase it is implicitly understood that the values that we assume for the uncertainties are all at the same level of confidence.

There are two nondimensionalized forms of Eq. (4.3) that are extremely useful in a planning phase uncertainty analysis. To obtain the first form, we divide each term in the equation by r^2. In each term on the right-hand side, we multiply by $(X_i/X_i)^2$, which of course is equal to 1. We then obtain

$$\frac{U_r^2}{r^2} = \left(\frac{X_1}{r}\frac{\partial r}{\partial X_1}\right)^2 \left(\frac{U_{X_1}}{X_1}\right)^2 + \left(\frac{X_2}{r}\frac{\partial r}{\partial X_2}\right)^2 \left(\frac{U_{X_2}}{X_2}\right)^2 + \cdots$$
$$+ \left(\frac{X_J}{r}\frac{\partial r}{\partial X_J}\right)^2 \left(\frac{U_{X_J}}{X_J}\right)^2 \qquad (4.6)$$

where U_r/r is the relative uncertainty of the result. The factors U_{X_i}/X_i are the relative uncertainties for each variable. In general, the relative uncertainties will be numbers less than 1, usually much less than 1.

We will call the factors in the parentheses that multiply the relative uncertainties of the variables *uncertainty magnification factors* (UMFs) and define them as

$$\text{UMF}_i = \frac{X_i}{r} \frac{\partial r}{\partial X_i} \quad (4.7)$$

The UMF for a given X_i indicates the influence of the uncertainty in that variable on the uncertainty in the result. A UMF value greater than 1 indicates that the influence of the uncertainty in the variable is magnified as it propagates through the data reduction equation into the result. A UMF value of less than 1 indicates that the influence of the uncertainty in the variable is diminished as it propagates through the data reduction equation into the result. Since the UMFs are squared in Eq. (4.6), their signs are of no importance. Thus we consider only the absolute values of the UMFs when we perform a general uncertainty analysis; in the past, the UMFs were sometimes called *normalized sensitivity coefficients*.

The second nondimensionalized form of Eq. (4.3) is found by dividing the equation by U_r^2 to obtain

$$1 = \frac{(\partial r/\partial X_1)^2 (U_{X_1})^2 + (\partial r/\partial X_2)^2 (U_{X_2})^2 + \cdots + (\partial r/\partial X_J)^2 (U_{X_J})^2}{U_r^2} \quad (4.8)$$

We then define an *uncertainty percentage contribution* (UPC) as

$$\text{UPC}_i = \frac{(\partial r/\partial X_i)^2 (U_{X_i})^2}{U_r^2} \times 100 = \frac{[(X_i/r)(\partial r/\partial X_i)]^2 (U_{X_i}/X_i)^2}{(U_r/r)^2} \times 100 \quad (4.9)$$

The UPC for a given X_i gives the percentage contribution of the uncertainty in that variable to the squared uncertainty in the result. Since the UPC of a variable includes the effects of both the UMF and the magnitude of the uncertainty of the variable, it is useful in the planning phase once we begin to make estimates of the uncertainties of the variables. This usually follows an initial analysis of the UMFs, for which no uncertainty estimates are necessary.

4-3 APPLICATION TO EXPERIMENT PLANNING (TSM)

4-3.1 Simple Case

Suppose that we want to answer the question: What is the density of air in a pressurized tank? Not having a density meter at hand, we consider what physical principles are available that will yield the value of the density if we determine the values of some other variables. If conditions are such that the ideal gas law applies, then we can determine the density of the air from

$$\rho = \frac{p}{RT} \quad (4.10)$$

if we are able to measure the air pressure and temperature and find the value of the gas constant for air from a reference source.

Equation (4.10) is in the general data reduction equation form of Eq. (4.2) since the density ρ is the result, so for this case Eq. (4.6) gives

$$\frac{U_\rho^2}{\rho^2} = \left(\frac{p}{\rho}\frac{\partial\rho}{\partial p}\right)^2\left(\frac{U_p}{p}\right)^2 + \left(\frac{R}{\rho}\frac{\partial\rho}{\partial R}\right)^2\left(\frac{U_R}{R}\right)^2 + \left(\frac{T}{\rho}\frac{\partial\rho}{\partial T}\right)^2\left(\frac{U_T}{T}\right)^2 \quad (4.11)$$

The UMFs are

$$\mathrm{UMF}_p = \frac{p}{\rho}\frac{\partial\rho}{\partial p} = \frac{p}{\rho RT} \quad (4.12)$$

$$\mathrm{UMF}_R = \frac{R}{\rho}\frac{\partial\rho}{\partial R} = \frac{R}{\rho}\left(\frac{-p}{R^2 T}\right) \quad (4.13)$$

$$\mathrm{UMF}_T = \frac{T}{\rho}\frac{\partial\rho}{\partial T} = \frac{T}{\rho}\left(\frac{-p}{RT^2}\right) \quad (4.14)$$

and substituting Eqs. (4.10), (4.12), (4.13), and (4.14) into Eq. (4.11) yields

$$\frac{U_\rho^2}{\rho^2} = \left(\frac{p}{RT}\frac{RT}{p}\right)^2\left(\frac{U_p}{p}\right)^2 + \left(\frac{-pR}{R^2 T}\frac{RT}{p}\right)^2\left(\frac{U_R}{R}\right)^2 + \left(\frac{-pT}{RT^2}\frac{RT}{p}\right)^2\left(\frac{U_T}{T}\right)^2 \quad (4.15)$$

or

$$\frac{U_\rho^2}{\rho^2} = \frac{U_p^2}{p^2} + \frac{U_R^2}{R^2} + \frac{U_T^2}{T^2} \quad (4.16)$$

Equation (4.16) relates the relative uncertainty in the experimental result, ρ, to the relative uncertainties in the measured variables and the gas constant. The values of universal constants (gas constant, π, Avogadro's number, etc.) are typically known with much greater accuracy than the measurements that are made in most experiments. It is common practice in the planning phase of an experiment to assume that the uncertainties in such quantities are negligible and to set them equal to zero. Assuming that U_R is zero, our uncertainty expression then becomes

$$\left(\frac{U_\rho}{\rho}\right)^2 = \left(\frac{U_p}{p}\right)^2 + \left(\frac{U_T}{T}\right)^2 \quad (4.17)$$

Obviously, in this simple example all UMFs are equal to 1. We are now in a position to answer many types of questions about the proposed experiment. Two questions that might be of interest are considered in the following examples.

Example 4.1 A pressurized air tank (Figure 4.1) is nominally at ambient temperature (25°C). How accurately can the density be determined if the temperature

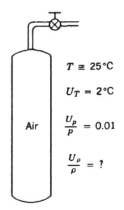

Figure 4.1 Sketch for Example 4.1.

is measured with an uncertainty of 2°C and the tank pressure is measured with a relative uncertainty of 1%?

Solution   The data reduction equation (4.10) is

$$\rho = \frac{p}{RT}$$

and the general uncertainty analysis expression (4.17) is

$$\left(\frac{U_\rho}{\rho}\right)^2 = \left(\frac{U_p}{p}\right)^2 + \left(\frac{U_T}{T}\right)^2$$

Values for the variables are

$$U_T = 2°C(= 2 \text{ K}) \qquad T = 25 + 273 = 298 \text{ K}$$

$$\frac{U_p}{p} = 0.01$$

since the 1% uncertainty in the pressure measurement is interpreted mathematically as

$$U_p = 0.01p$$

Substitution of the numerical values into (4.17) yields

$$\left(\frac{U_\rho}{\rho}\right)^2 = (0.01)^2 + \left(\frac{2 \text{ K}}{298 \text{ K}}\right)^2$$

$$= 1.0 \times 10^{-4} + 0.45 \times 10^{-4} = 1.45 \times 10^{-4}$$

and taking the square root gives the uncertainty in the density

$$\frac{U_\rho}{\rho} = 0.012 \quad \text{or} \quad 1.2\%$$

Example 4.2   For the physical situation in Example 4.1, suppose that the density determination must be within 0.5% to be useful. If the temperature measurement can be made to within 1°C, how accurate must the pressure measurement be? (See Figure 4.2.)

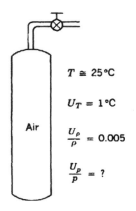

Figure 4.2   Sketch for Example 4.2.

Solution   Again Eq. (4.17) is used,

$$\left(\frac{U_\rho}{\rho}\right)^2 = \left(\frac{U_p}{p}\right)^2 + \left(\frac{U_T}{T}\right)^2$$

with numerical values of

$$U_T = 1°C (= 1 \text{ K}) \qquad T = 25 + 273 = 298 \text{ K}$$

$$\frac{U_\rho}{\rho} = 0.005$$

so that we obtain

$$(0.005)^2 = \left(\frac{U_p}{p}\right)^2 + \left(\frac{1 \text{ K}}{298 \text{ K}}\right)^2$$

or

$$\left(\frac{U_p}{p}\right)^2 = 1.37 \times 10^{-5} \qquad \frac{U_p}{p} = 0.0037 \quad \text{or} \quad 0.37\%$$

The pressure measurement would have to be made with an uncertainty of less than 0.37% for the density determination to meet the specifications.

4-3.2 Special Functional Form

A very useful specific form of Eq. (4.6) is obtained when the data reduction equation (4.2) has the form

$$r = kX_1^a X_2^b X_3^c \cdots \tag{4.18}$$

where the exponents may be positive or negative constants and k is a constant. Application of Eq. (4.6) to the relationship of (4.18) yields

$$\left(\frac{U_r}{r}\right)^2 = a^2\left(\frac{U_{X_1}}{X_1}\right)^2 + b^2\left(\frac{U_{X_2}}{X_2}\right)^2 + c^2\left(\frac{U_{X_3}}{X_3}\right)^2 + \cdots \tag{4.19}$$

A data reduction equation of the form (4.18) is thus especially easy to work with, as the result of the uncertainty analysis may be written down by inspection in the form (4.19) without any partial differentiation and subsequent algebraic manipulation. One must remember, however, that the X_i's in Eq. (4.18) represent variables that are directly measured. Thus

$$W = I^2 R \tag{4.20}$$

is of the requisite form when the current I and the resistance R are measured, but

$$Y = Z \sin\theta \tag{4.21}$$

is not if Z and θ are the measured variables, and

$$V = Z(p_2 - p_1)^{1/2} \tag{4.22}$$

is not if p_2 and p_1 are measured separately. If the difference $(p_2 - p_1)$ is measured directly, however, then (4.22) is in a form where (4.19) can be applied.

For the special functional form, the exponents correspond to the UMFs. Thus the influence of uncertainties in variables raised to a power with magnitude greater than 1 is magnified. Conversely, the influence of the uncertainties in variables raised to a power with magnitude less than 1 is diminished.

Example 4.3   We again plan to use the ideal gas law to determine the density of a gas. The density determination must be accurate to within 0.5%, and we can measure the temperature (nominally 25°C) to within 1°C. What is the maximum allowable uncertainty in the pressure measurement? (See Figure 4.3.)

Solution   The ideal gas law

$$p = \rho R T$$

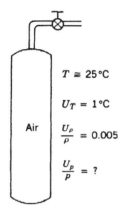

Figure 4.3   Sketch for Example 4.3.

is again our data reduction equation. This time we note that this is in the special form of Eq. (4.18), and assuming zero uncertainty in the gas constant, we can write down by inspection

$$\left(\frac{U_p}{p}\right)^2 = \left(\frac{U_\rho}{\rho}\right)^2 + \left(\frac{U_T}{T}\right)^2$$

and substituting in the given information we find that

$$\left(\frac{U_p}{p}\right)^2 = (0.005)^2 + \left(\frac{1\text{ K}}{298\text{ K}}\right)^2$$

and the required uncertainty in the pressure measurement is

$$\frac{U_p}{p} = 0.0061 \quad \text{or} \quad 0.61\%$$

BUT WAIT! THIS IS INCORRECT. We did not solve for the experimental result (ρ) before we did the uncertainty analysis. To use Eq. (4.19), one must solve for the result first, so

$$\rho = p^1 R^{-1} T^{-1}$$

and using (4.19) and assuming U_R negligible yield

$$\left(\frac{U_\rho}{\rho}\right)^2 = \left(\frac{U_p}{p}\right)^2 + \left(\frac{U_T}{T}\right)^2$$

Substituting in the given information, we find that

$$(0.005)^2 = \left(\frac{U_p}{p}\right)^2 + \left(\frac{1\ \text{K}}{298\ \text{K}}\right)^2$$

so that

$$\left(\frac{U_p}{p}\right)^2 = (0.005)^2 - (0.0034)^2$$

and the required uncertainty in the pressure measurement is

$$\frac{U_p}{p} = 0.0037 \quad \text{or} \quad 0.37\%$$

which agrees with the result found in Example 4.2. Note that the required uncertainty was erroneously found to be 1.6 times the correct value when the incorrect analysis was done.

This example is intended to show how easily a mistake can be made when the data reduction equation is of the special form and the uncertainty expression is written down by inspection. *It is imperative that the data reduction equation be solved for the experimental result before the uncertainty analysis is begun.*

Example 4.4   The power W drawn by a resistive load in a direct-current (dc) circuit (Figure 4.4) can be determined by measuring the voltage E and current I and using

$$W = EI$$

It can also be determined by measuring the resistance R and the current and using

$$W = I^2 R$$

If I, E, and R can all be measured with approximately equal uncertainties (on a relative or percentage basis), which method of power determination is "best"?

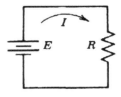

Figure 4.4   Sketch for Example 4.4.

Solution   All other things (cost, difficulty of measurement, etc.) being equal, the best method will be the one that produces a result with the least uncertainty. Both data reduction expressions are in the special form of Eq. (4.18). For $W = EI$,

$$\left(\frac{U_W}{W}\right)^2 = \left(\frac{U_E}{E}\right)^2 + \left(\frac{U_I}{I}\right)^2$$

and all UMFs are 1. For $W = I^2 R$,

$$\left(\frac{U_W}{W}\right)^2 = (2)^2\left(\frac{U_I}{I}\right)^2 + \left(\frac{U_R}{R}\right)^2$$

where the UMF for I is 2, indicating that its uncertainty is magnified. If the E, I, and R measurements have equal percentage uncertainties, the second method produces a result with an uncertainty 1.58 times that of the first method. This can be verified by substitution of any assumed percentage value of uncertainty for the measured variables.

This example is intended to emphasize the usefulness of uncertainty analysis in the planning phase of an experiment and the generality of the conclusions that can often be drawn. Note the conclusion we were able to make in this case without knowing the level of uncertainty in the measurements, what instruments might be available, or any other details.

Example 4.5   It is proposed that the shear modulus M_S be determined for an alloy by measuring the angular deformation θ produced when a torque T is applied to a cylindrical rod of the alloy with radius R and length L. The expression relating these variables [1] is

$$\theta = \frac{2LT}{\pi R^4 M_S}$$

We wish to examine the sensitivity of the experimental result to the uncertainties in the variables that must be measured before we proceed with a detailed experimental design.

Solution   The physical situation shown in Figure 4.5 (where torque T is given by aF) is described by the data reduction equation for the shear modulus,

$$M_S = \frac{2LaF}{\pi R^4 \theta}$$

Noting that this is in the special form of Eq. (4.18) and assuming as usual that the factors 2 and π have zero uncertainties, general uncertainty analysis yields

$$\left(\frac{U_{M_S}}{M_S}\right)^2 = \left(\frac{U_L}{L}\right)^2 + \left(\frac{U_a}{a}\right)^2 + \left(\frac{U_F}{F}\right)^2 + 16\left(\frac{U_R}{R}\right)^2 + \left(\frac{U_\theta}{\theta}\right)^2$$

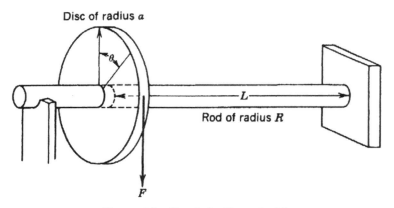

Figure 4.5   Sketch for Example 4.5.

This shows the tremendous sensitivity of the uncertainty in the result to the uncertainty in variables whose exponents (UMFs) are large. In this case, if all variables are measured with 1% uncertainty, the uncertainty in M_S is 4.5%, whereas it still would be 4% if U_L, U_a, U_F, and U_θ were all assumed zero.

This example is intended to illustrate the utility of uncertainty analysis in identifying the measurements with which special care must be taken. Since this information can be determined in the planning phase of an experiment, it can be incorporated into the design of an experiment from the very beginning.

Note that the value assigned to the uncertainty in R should include the effects of the variation of radius along the length of the rod from the nominal value of R. The results of the general uncertainty analysis can thus guide the specification of the tolerances to which the rod must be manufactured for the experiment. If the uncertainty with which the shear modulus must be determined is known, an estimate of the maximum permissible tolerances on the rod dimensions can be made.

4-4 USING TSM UNCERTAINTY ANALYSIS IN PLANNING AN EXPERIMENT

In the preceding sections of this chapter we discussed general uncertainty analysis and some initial examples of its application. It is worthwhile at this point to step back from the details and reemphasize how this technique fits into the experimental process and, particularly, how useful it is in the planning phase of experiments.

There is a question to which an experimental answer is sought. The planning phase of an experiment is the period during which the physical phenomena that might lead to such an answer—the potential data reduction equations—are considered. This is also sometimes called the preliminary design stage of an experiment.

It is important even at this stage that the allowable uncertainty in the result be known. This seems to be a point that bothers many, particularly those who are relative novices at experimentation. However, it is ridiculous to embark on an experimental effort without considering the degree of goodness necessary in the result. Here we are talking about whether the result needs to be known within 0.1, 1, 10, or 50%, not whether 1.3 versus 1.4% is necessary. The degree of uncertainty allowable in the result can usually be estimated fairly well by considering the use that will be made of the result.

In the planning phase, general uncertainty analysis is used to investigate whether it is feasible to pursue a particular experiment, which measurements might be more important than others, which techniques might give better results than others, and similar matters. We are attempting to understand the behavior of the experimental result as a function of the uncertainties in the measured variables and to use this to our advantage, if possible. At this stage we are not worrying about whether brand A or brand B instrument should be purchased or whether a temperature should be measured with a thermometer, thermocouple, or thermistor.

In using general uncertainty analysis, uncertainties must be assigned for the variables that will be measured and properties that typically will be found from reference sources. In years of teaching, the authors have observed an almost universal reluctance to estimate uncertainties. There seems to be a feeling that "out there" somewhere is an expert who knows what all these uncertainties *really* are and who will suddenly appear and scoff at any estimates we make. This is nonsense! Uncertainty analysis in the planning phase can be viewed as a spreadsheet that allows us to play "what if" before making decisions on the specifics of an experimental design.

If there is not an experience base for estimating the uncertainties, a parametric analysis using a range of values can be made. There are no penalties at this stage for making ridiculous estimates—if we want to see the effect on a result assuming a temperature measurement uncertainty of 0.00001°F, we are free to do so.

The primary point to be made is that a general uncertainty analysis should be used in the very initial stages of an experimental program. The information and insight gained are far out of proportion to the small amount of time the analysis takes, and parametric analysis using a range of assumed values for the uncertainties is perfectly acceptable.

In the following sections of this chapter, comprehensive examples of the use of general uncertainty analysis in the planning phase of two experimental programs are presented. The time that the reader spends in working through these examples in detail will be well spent.

4-5 EXAMPLE: ANALYSIS OF PROPOSED PARTICULATE MEASURING SYSTEM

4-5.1 The Problem

A manufacturer needs to determine the solid particulate concentration in the exhaust gases from one of its process units. Monitoring the concentration will allow immediate recognition of unexpected and undesired changes in the operating condition of the unit. The measurement is also necessary to determine and monitor compliance with air pollution regulations. The measurement needs to be accurate within about 5%, with a 10% or greater inaccuracy being unacceptable.

A representative of an instrument company has submitted a presentation proposing that the manufacturer purchase and install the model LT1000 laser transmissometer system to measure the projected area concentration of the particulate matter in the stack exhaust. He states that the system will meet the 5% uncertainty requirement. Management has directed the engineering department to evaluate the salesman's proposal and recommend whether the system should be purchased.

4-5.2 Proposed Measurement Technique and System

A schematic of the laser transmissometer system is shown in Figure 4.6. A laser beam passes through the exhaust gas stream and impinges on a transducer that measures the beam intensity. The physical process is described [2] by the expression

$$T = \frac{I}{I_0} = e^{-CEL} \qquad (4.23)$$

In this expression, I_0 is the intensity of the light beam exiting the laser, I is the transmitted intensity after the beam has passed through the scattering and absorbing medium of thickness L, and T is the fractional transmitted intensity (the transmittance). The projected area concentration C is the projected area of the particulates per unit volume of the medium.

The extinction coefficient E is a function of the wavelength of the laser light beam, the distribution of sizes and shapes of the particles, and the indices of refraction of the particles. It approaches an asymptotic value of 2.0 for certain optical conditions. The instrument company states in its proposal that based on experience with exhaust stream measurements in similar plants, the value of E is expected to be "within a couple of percent of 2.0" over the range of operating conditions expected.

The system is adjusted so that when the laser beam travels through the stack with no exhaust flow, the power meter output reads 1.000. When there is flow through the stack, therefore, the output of the power meter corresponds directly to the value of the transmittance T. The instrument manufacturer specifies that the power meter is accurate to 1% of reading or better.

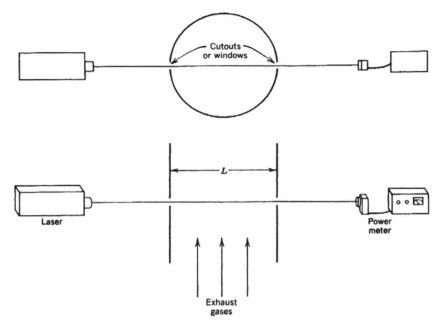

Figure 4.6   Schematic of a laser transmissometer system for monitoring particulate concentrations in exhaust gases.

4-5.3 Analysis of Proposed Experiment

We have the assignment of deciding whether the proposed experimental approach will answer the question of interest (What is the projected area concentration of the particulates in the exhaust?) acceptably (with an uncertainty of about 5% of less). Although we may never have touched a laser or seen the particular exhaust stack in question, general uncertainty analysis gives us a logical, proven technique for evaluating the proposed experiment in the planning stage.

The data reduction equation (4.23) is

$$T = e^{-CEL}$$

However, the experimental result is C, the particulate projected area concentration. Taking the natural logarithm of Eq. (4.23) and solving for C yield

$$C = -\frac{\ln T}{EL} \qquad (4.24)$$

Since T, and not $\ln T$, is measured, Eq. (4.24) is not in the special product form, and the proper uncertainty analysis expression to use is Eq. (4.6). For this case

$$\left(\frac{U_C}{C}\right)^2 = \left[\left(\frac{T}{C}\right)\frac{\partial C}{\partial T}\frac{U_T}{T}\right]^2 + \left[\left(\frac{E}{C}\right)\frac{\partial C}{\partial E}\frac{U_E}{E}\right]^2 + \left[\left(\frac{L}{C}\right)\frac{\partial C}{\partial L}\frac{U_L}{L}\right]^2 \qquad (4.25)$$

Proceeding with the partial differentiation gives

$$\text{UMF}_T = \frac{T}{C}\frac{\partial C}{\partial T} = \frac{-ELT}{\ln T}\left(-\frac{1}{EL}\right)\frac{1}{T} = \frac{1}{\ln T} \tag{4.26}$$

$$\text{UMF}_E = \frac{E}{C}\frac{\partial C}{\partial E} = \frac{-E^2 L}{\ln T}\left(-\frac{\ln T}{L}\right)\left(-\frac{1}{E^2}\right) = -1 \tag{4.27}$$

$$\text{UMF}_L = \frac{L}{C}\frac{\partial C}{\partial L} = \frac{-EL^2}{\ln T}\left(-\frac{\ln T}{E}\right)\left(-\frac{1}{L^2}\right) = -1 \tag{4.28}$$

and substitution into (4.25) yields

$$\left(\frac{U_C}{C}\right)^2 = \left(\frac{1}{\ln T}\right)^2\left(\frac{U_T}{T}\right)^2 + \left(\frac{U_E}{E}\right)^2 + \left(\frac{U_L}{L}\right)^2 \tag{4.29}$$

This is the desired expression relating the uncertainty in the result to the uncertainties in the measured quantities T and L and the assumed value of E.

Equation (4.29) shows that the uncertainty in C depends not only on the uncertainties in L, T, and E but also on the value of the transmittance itself since the UMFs for E and L are 1 but the UMF for T is a function of T and grows without bound as T approaches 1. The uncertainty in C will thus vary with operating conditions even if the uncertainties in L, T, and E are all constant. In situations such as this, a parametric study is indicated.

In considering the physical situation, we can see that the transmittance T will vary between the limiting values 0 and 1. When $I = I_0$, $T = 1.0$, that is, when there are no particles in the flow through the stack. In the other limit, $T = 0$ when the flow is opaque and no light makes it through the flow. If we investigate the behavior of the uncertainty in the result for values of T between 0 and 1, the entire possible range of operating conditions will have been covered. To calculate values of U_C/C, we must first estimate values of the uncertainties in T, E, and L.

This is the point at which those unfamiliar with uncertainty analysis and inexperienced in either experimentation or the particular technique being considered typically feel uneasy. A reluctance to estimate uncertainty values seems to be a part of human nature. Such reluctance can be overcome by choosing a range of uncertainties that will almost certainly bracket the "true" uncertainties. In this case we would calculate U_C/C for a set of T values between 0 and 1 for assumed uncertainties in T, E, and L of 0.1, 1, 10, and 50% if we wished.

In situations such as this, however, it is often quite revealing to initially choose all the uncertainties the same and proceed with the parametric study. If we choose all uncertainties as 1%, then

$$\left(\frac{U_C}{C}\right)^2 = \left(\frac{1}{\ln T}\right)^2(0.01)^2 + (0.01)^2 + (0.01)^2$$

$$= \left(\frac{1}{\ln T}\right)^2(10^{-4}) + 2.0 \times 10^{-4} \tag{4.30}$$

We note that, as $T \to 0$,

$$\ln T \to -\infty$$

and the first term on the right-hand side approaches zero. Conversely, as $T \to 1$,

$$\ln T \to 0$$

and the first term on the right-hand side grows without bound. The uncertainty percentage contribution (UPC) for T thus varies from 0 to 100% as T goes from 0 to 1. The behavior of Eq. (4.30) over the entire range of T from 0 to 1 is shown in Figure 4.7.

4-5.4 Implications of Uncertainty Analysis Results

The results of the uncertainty analysis reveal a behavior that is certainly not obvious simply by looking at the expressions [Eqs. (4.23) and (4.24)] that model this physical phenomenon. Even if all measurements are made with 1% uncertainty, the uncertainty in the experimental result can be 10%, 20%, 100%, or greater.

It should be noted that this behavior does not occur at operating conditions that are unlikely to be encountered. To the contrary, this behavior is observed in the range of operating conditions that one would anticipate being of interest. For there to be transmittances of 0.7, 0.6, or less, over a relatively short distance such as a typical stack diameter, the solid particle concentration in the exhaust gases would have to be tremendous. It would seem likely, therefore, that measurements with T less than 0.7 or 0.6 would occur at conditions unacceptable from a

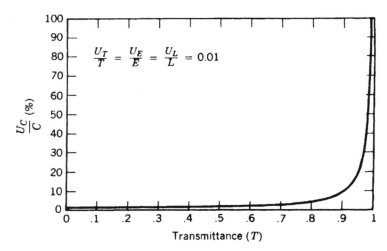

Figure 4.7   Uncertainty in the experimental result as a function of the transmittance.

pollution standpoint. We can conclude, then, that the technique as proposed is not acceptable. It could not be expected to yield results with uncertainty of 5 to 10% or less over the range of operating conditions one would expect to encounter.

It should be noted that this conclusion was reached without our having to investigate the details of the dependence of E, the extinction coefficient, on the nature of the particles. The results of the uncertainty analysis showed unacceptable behavior of the uncertainty in the experimental result even with an optimistic assumption of 1% uncertainty in the value of E.

4-5.5 Design Changes Indicated by Uncertainty Analysis

In looking at the uncertainty analysis results plotted in Figure 4.7, we see that there would be a chance that the technique would be acceptable if the transmittance were always 0.8 or less. Since we are doing an uncertainty analysis in the planning phase of an experiment, we are free to play "what if" with almost no restrictions. In fact, playing "what if" at this stage of an experimental design should be encouraged.

Consider expression (4.23) for the transmittance:

$$T = e^{-CEL}$$

For a given set of operating conditions, the characteristics of the exhaust flow are fixed and therefore C and E are fixed. The only way to cause a decrease in T, then, is to increase L, the path length of the light beam through the exhaust stream. Rather than recommend that a new, larger diameter stack be constructed (!), we might recommend that two mirrors on adjustable mounts be purchased and used, as shown in Figure 4.8.

The two additional mirrors allow multiple passes of the beam through the stack, thus increasing the path length L, decreasing the overall transmittance, and decreasing the uncertainty in the measurement of C. For example, consider the case in which all measurements are made with 1% uncertainty and

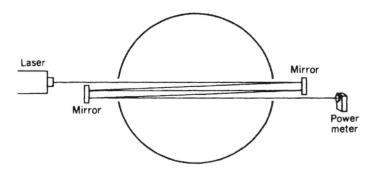

Figure 4.8   Multiple-pass beam arrangement for the transmissometer.

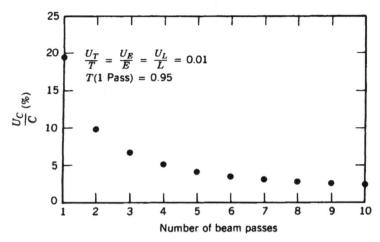

Figure 4.9   Uncertainty in the experimental result as a function of the number of beam passes.

the transmittance for one pass of the beam through the stack is 0.95. The effect of additional passes on the transmittance can be calculated from $T(n \text{ passes}) = (0.95)^n$. The behavior of the uncertainty in the result is shown in Figure 4.9 as a function of the number of beam passes through the stack.

The results of the analysis above indicate that the proposed system, with modifications, might be able to meet the requirements. Additional factors, such as the effect of laser output (I_0) variation with time and the probable behavior of E, should be investigated if the implementation of the technique is to be considered further [3, 4].

4-6 EXAMPLE: ANALYSIS OF PROPOSED HEAT TRANSFER EXPERIMENT

In this section we use general uncertainty analysis to investigate the suitability of two techniques that might be used to determine the convective heat transfer coefficient for a circular cylinder immersed in an airflow. An extensive analysis and discussion of this case has been presented by Moffat [5].

4-6.1 The Problem

A client of our research laboratory needs to determine the behavior of the convective heat transfer from a circular cylinder of finite length that will be immersed in an airflow whose velocity might range from 1 to 100 m/s. The physical situation and nomenclature are shown in Figure 4.10.

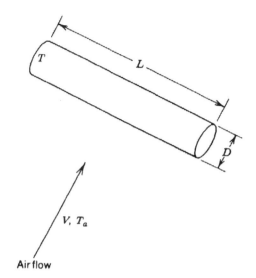

Figure 4.10 Finite-length circular cylinder in crossflow.

The convective heat transfer coefficient h is defined by the relationship

$$q = hA(T - T_a) \qquad (4.31)$$

where q is the rate of convective heat transfer from the cylinder to the air stream, T is the temperature of the surface of the cylinder, T_a is the temperature of the oncoming air, and A is the surface area, given in this case by

$$A = \pi DL \qquad (4.32)$$

For T, T_a, and A remaining constant, h (and hence q) increases with increasing air velocity V.

The cylindrical tube to be tested is 0.152 m (6.0 in.) long with a diameter of 0.0254 m (1.0 in.) and a wall thickness of 1.27 mm (0.050 in.). The primary case of interest is for an air temperature of about 25°C with the cylinder about 20°C above that. The cylinder is made of aluminum, for which the specific heat has been found from a table [6] as 0.903 kJ/kg·°C at 300 K. The mass of the tube has been measured as 0.020 kg. The client wants to know h within 5% and must be assured that h can be determined within 10% before funding for the experiment will be allocated.

4-6.2 Two Proposed Experimental Techniques

Two techniques have been suggested for determining the heat transfer coefficient. The first is a steady-state technique, and the second is a transient technique.

Steady-State Technique. As shown schematically in Figure 4.10, the cylinder is mounted in a 25°C airstream of velocity V. Sufficient electric power is supplied to a resistance heater in the cylindrical tube so that the surface temperature reaches a steady-state value of about 45°C. We now apply the first law of thermodynamics to a control volume whose surface lies infinitesimally outside the surface of the cylinder. Assuming the radiative and conductive heat transfer losses to be negligible for purposes of this initial analysis, the energy balance yields

$$\text{Power in} = \text{rate of convected energy out} \qquad (4.33)$$

or

$$W = h(\pi DL)(T - T_a) \qquad (4.34)$$

Solving for the experimental result,

$$h = \frac{W}{\pi DL(T - T_a)} \qquad (4.35)$$

For a given air velocity, then, the convective heat transfer coefficient can be determined by measuring the electrical power W, cylinder diameter D, length L, and temperatures of the cylinder surface and the airstream.

Transient Technique. This approach uses the characteristics of the response of a first-order system to a step change in the input to determine the heat transfer coefficient. The experimental setup is similar to that for the steady-state case. The cylinder is heated until it reaches a temperature T_1, say, and then the power is switched off. The cylinder is then cooled by the airstream, and its temperature T will eventually reach T_a. The temperature-versus-time behavior is shown in Figure 4.11.

The time taken for the cylinder temperature T to change 63.2% of the way from T_1 to T_a is equal to the time constant τ of the system, where τ is related to the convective heat transfer coefficient by

$$\tau = \frac{Mc}{h(\pi DL)} \qquad (4.36)$$

where M is the mass of the cylinder and c is the specific heat of the cylinder.

Solving for the experimental result, we have

$$h = \frac{Mc}{\pi DL\tau} \qquad (4.37)$$

For a given air velocity, the heat transfer coefficient can be determined by measuring the mass, diameter, and length of the cylinder, finding the specific heat of the cylinder, and measuring a time corresponding to τ.

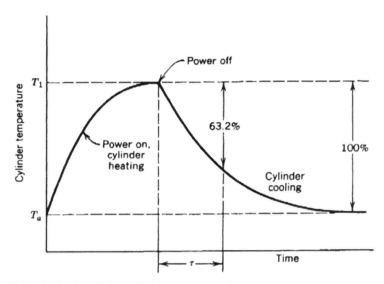

Figure 4.11   Behavior of the cylinder temperature in a test using the transient technique.

In the above, radiative and conductive heat transfer losses are assumed negligible, h and c are assumed constant during the cooling process, and the cylinder is assumed to have negligible internal resistance (each point in the cylinder is at the same temperature T as every other point in the cylinder). This last assumption holds true for Biot numbers less than $\frac{1}{10}$ [6].

To determine whether the techniques can produce results with acceptable uncertainty, if one technique has advantages over the other, or if there is any behavior in the uncertainty in the result that is not obvious from Eqs. (4.35) and (4.37), a general uncertainty analysis should be performed for each of the techniques proposed.

4-6.3 General Uncertainty Analysis: Steady-State Technique

The data reduction expression for the steady-state technique is given by Eq. (4.35):

$$h = \frac{W}{\pi DL(T - T_a)}$$

If the temperatures are measured separately, this expression is not in the special product form [Eq. (4.18)] and the general uncertainty analysis must be done using Eq. (4.6). Leaving the details as an exercise for the reader, the analysis yields

$$\left(\frac{U_h}{h}\right)^2 = \left(\frac{U_W}{W}\right)^2 + \left(\frac{U_D}{D}\right)^2 + \left(\frac{U_L}{L}\right)^2 + \left(\frac{U_T}{T - T_a}\right)^2 + \left(\frac{U_{T_a}}{T - T_a}\right)^2 \quad (4.38)$$

For the present analysis, assume that the temperature difference $(T - T_a)$ will be measured directly, perhaps with a differential thermocouple circuit. In this case, Eq. (4.35) is in the special product form, and we can write by inspection

$$\left(\frac{U_h}{h}\right)^2 = \left(\frac{U_W}{W}\right)^2 + \left(\frac{U_D}{D}\right)^2 + \left(\frac{U_L}{L}\right)^2 + \left(\frac{U_{\Delta T}}{\Delta T}\right)^2 \qquad (4.39)$$

where $\Delta T = T - T_a$.

At this point one might be tempted simply to estimate all the uncertainties as 1%, and Eq. (4.39) would then yield the uncertainty in the result, h, as 2%. This example, however, provides us with an opportunity to discuss several important practical points about using uncertainty analysis in planning experiments.

When an expression for the fractional uncertainty in a temperature (U_T/T) is encountered (as in Examples 4.1, 4.2, and 4.3), the temperature T *must* be expressed in absolute units of degrees Kelvin or Rankine and not in degrees Celsius or Fahrenheit. Similarly, the pressure p in U_p/p must be absolute, not gauge. However, an absolute uncertainty in a quantity has the dimensions of a difference of two values of that quantity, and for U_T this means the units are Celsius degrees (which are equal to Kelvin degrees) or Fahrenheit degrees (which are equal to Rankine degrees). Thus, if someone says that he or she can measure a temperature of 27°C with 1% relative uncertainty, U_T is 3°C (which is 0.01 times 300 K) and not 0.27°C. In general, percentage specifications for uncertainties in temperature measurements are easily misinterpreted and should be avoided.

In Eq. (4.39), we are working with a temperature difference and have the quantity $U_{\Delta T}/\Delta T$. If we estimate this as 1%, we are saying that the uncertainty is 0.2°C when $\Delta T = 20°C$, the nominal value planned for the steady-state tests. However, if for some reason a test were run with $\Delta T = 5°C$, our 1% specification indicates that the uncertainty is 0.05°C, which is certainly overly optimistic for most experimental programs. The point to be emphasized here is that it is easier to estimate uncertainties on a relative (or percentage) basis, but if a range of values of a variable is of interest or is likely to be encountered, estimating the uncertainty in that variable on an absolute basis gives less chance for misinterpretation of the behavior of the experiment.

The uncertainty in the power measurement in the steady-state technique is just such a case. Since we are interested in determining h for a range of air velocities, we would anticipate encountering a range of values for h. From Eq. (4.35) we would then expect a range of values for the power W. If we estimate the uncertainty in the measurement of W as 2%, say, then U_W is 2 W when $W = 100$ W and 0.02 W when $W = 1$ W. The authors have found that it is, in general, productive to investigate a range of assumed absolute uncertainties in the planning phase of an experiment—that this gives the best opportunity to observe the behavior of the uncertainty in the result. In this case, for instance, one might want to check the behavior of the uncertainty in h for several assumed values of the uncertainty in W.

From Eq. (4.39) we can see that if we estimate a value for U_W, we must have a numerical estimate of W also before U_h/h can be calculated. The amount of power W necessary to maintain a ΔT value of 20°C is dependent on the value of h [see Eq. (4.35)], so we must estimate the range of values we expect to encounter in the result h before we can proceed with the uncertainty analysis. Considering the range of air velocities of interest and published results for convective heat transfer from infinitely long cylinders, it is reasonable to consider a range of h values from 10 to 1000 W/m² · °C (about 1.8 to 180 Btu/hr · ft² · °F). Putting numerical values into Eq. (4.35), this corresponds to values of input power W from 2.4 to 243 W.

We now need to estimate the uncertainties in D, L, ΔT, and W. Using

$$U_D = 0.025 \text{ mm } (0.001 \text{ in.})$$

$$U_L = 0.25 \text{ mm } (0.010 \text{ in.})$$

should be of the right order for the measurements of the cylinder dimensions. For the differential temperature measurement, let us initially consider

$$U_{\Delta T} = 0.25°\text{C and } 0.5°\text{C}$$

For the power measurement, which will range up to 243 W at the highest h value assumed, estimates of

$$U_W = 0.5 \text{ W and } 1 \text{ W}$$

correspond to 0.2 and 0.4% of full scale on a meter with a 250-W range. Substitution into Eq. (4.39) using the lower estimates gives

$$
\begin{array}{cccc}
W: & D: & L: & \Delta T:
\end{array}
$$

$$
\left(\frac{U_h}{h}\right)^2 = \left(\frac{0.5 \text{ W}}{W}\right)^2 + \left(\frac{0.025 \text{ mm}}{25.4 \text{ mm}}\right)^2 + \left(\frac{0.25 \text{ mm}}{152 \text{ mm}}\right)^2 + \left(\frac{0.25°\text{C}}{20°\text{C}}\right)^2
$$

$$
= \left(\frac{0.5 \text{ W}}{W}\right)^2 + (0.01 \times 10^{-4}) + (0.03 \times 10^{-4}) + (1.6 \times 10^{-4})
$$

$$
= \left(\frac{0.5 \text{ W}}{W}\right)^2 + (1.64 \times 10^{-4}) \tag{4.40}
$$

It is immediately apparent that the effect of the uncertainties in the cylinder dimensions is negligible relative to the influence of $U_{\Delta T}/\Delta T$ and that the relative influence of the uncertainty in the power measurement will be greater at low power and less at high power. To obtain numerical values from Eq. (4.40), we use Eq. (4.34):

$$W = h(\pi DL)(T - T_a)$$

which gives, upon use of the nominal values of D, L, and ΔT,

$$W = (0.01213 \text{ m}^2)(20°\text{C})h = 0.2426h \qquad (4.41)$$

where W is in watts and h in $\text{W/m}^2 \cdot °\text{C}$.

We can now choose values of h throughout the range of interest, use Eq. (4.41) to obtain the corresponding values of power W, and then use Eq. (4.40) to determine the uncertainty in the experimental result h. This has been done for the measurement uncertainties previously assumed, and the results of the analysis are shown in Table 4.1 and are plotted in Figure 4.12.

The results in the figure show that the assumed values for the uncertainties in the differential temperature measurement will satisfy the requirements and could

Table 4.1 Results of General Uncertainty Analysis for Steady-State Technique

		$U_W = 0.5$ W		$U_W = 1.0$ W	
$h(\text{W/m}^2 \cdot °\text{C})$	W (W)	$U_{\Delta T} =$ 0.25°C	$U_{\Delta T} =$ 0.50°C	$U_{\Delta T} =$ 0.25°C	$U_{\Delta T} =$ 0.50°C
10	2.4	20.8	20.8	41.7	41.7
20	4.9	10.3	10.5	20.5	20.6
50	12	4.4	4.9	8.4	8.7
100	24	2.4	3.3	4.4	4.9
200	49	1.6	2.7	2.4	3.2
500	121	1.3	2.6	1.5	2.7
1000	243	1.3	2.5	1.3	2.6

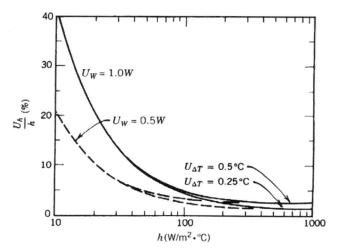

Figure 4.12 Uncertainty analysis results for the steady-state technique.

even be relaxed somewhat. The uncertainties of 0.5 and 1 W in the power measurement are adequate at higher values of h but do not satisfy the requirements
at low and moderate h values. The uncertainty in the power measurement is the
dominant uncertainty (for the assumptions made in this analysis) in the low and
moderate h range and would have to be reduced below 0.5 W for the requirement
of about 5% uncertainty in h to be met.

4-6.4 General Uncertainty Analysis: Transient Technique

The data reduction expression for the transient technique is given by Eq. (4.37):

$$h = \frac{Mc}{\pi D L \tau}$$

Then

$$\left(\frac{U_h}{h}\right)^2 = \left(\frac{U_\tau}{\tau}\right)^2 + \left(\frac{U_D}{D}\right)^2 + \left(\frac{U_L}{L}\right)^2 + \left(\frac{U_M}{M}\right)^2 + \left(\frac{U_c}{c}\right)^2 \qquad (4.42)$$

The uncertainties in D and L have been estimated in the analysis for the
steady-state case, and the same values will be used in this analysis. We should
be able to measure the mass of the cylinder to within, say, 0.1 g, so we estimate

$$U_M = 0.0001 \text{ kg}$$

Rather than return to the literature to find some of the original data on the specific
heat of aluminum, let us estimate the uncertainty in c as 1% for the planning
phase analysis. If the conclusions based on the results of our initial analysis turn
out to be a strong function of this estimate, we can then spend the time to obtain
a more defensible estimate. For the uncertainty in the measurement of the time
corresponding to the time constant, it makes sense to see the effect of a range of
values. For the initial analysis, we choose

$$U_\tau = 5, 10, \text{ and } 20 \text{ ms}$$

Substituting into Eq. (4.42) we find that

$$\begin{array}{ccccc} \tau. & D: & L: & M: & c: \end{array}$$

$$\left(\frac{U_h}{h}\right)^2 = \left(\frac{U_\tau}{\tau}\right)^2 + \left(\frac{0.025 \text{ mm}}{25.4 \text{ mm}}\right)^2 + \left(\frac{0.25 \text{ mm}}{152 \text{ mm}}\right)^2 + \left(\frac{0.0001 \text{ kg}}{0.020 \text{ kg}}\right)^2 + (0.01)^2$$

$$= \left(\frac{U_\tau}{\tau}\right)^2 + (0.01 \times 10^{-4}) + (0.03 \times 10^{-4})$$

$$\quad + (0.25 \times 10^{-4}) + (1 \times 10^{-4})$$

$$= \left(\frac{U_\tau}{\tau}\right)^2 + (1.29 \times 10^{-4}) \qquad (4.43)$$

The value of τ corresponding to an assumed value h can be found by substituting the values of M, c, D, and L into Eq. (4.36):

$$\tau = \frac{Mc}{h(\pi DL)}$$

to obtain

$$\tau = \frac{148.9}{h} \tag{4.44}$$

where τ is in seconds and h is in W/m$^2 \cdot {}^\circ$C.

In a manner similar to that used in the steady-state analysis, we can now choose a value of h, determine the corresponding value of τ from Eq. (4.44), and substitute into Eq. (4.43) to obtain the uncertainty in the result h. This has been done for the range of values assumed for U_τ, and the results are shown in Table 4.2 and plotted in Figure 4.13.

Table 4.2   Results of General Uncertainty Analysis for Transient Technique

h(W/m$^2 \cdot {}^\circ$C)	τ(s)	$U_\tau = 0.005$ s	$U_\tau = 0.010$ s	$U_\tau = 0.020$ s
		\multicolumn{3}{c}{$U_h/h(\%)$}		
10	14.9	1.1	1.1	1.3
20	7.4	1.1	1.1	1.3
50	3.0	1.2	1.2	1.4
100	1.5	1.2	1.3	1.8
200	0.74	1.3	1.8	3.0
500	0.30	2.0	3.5	6.8
1000	0.15	3.5	6.8	13.4

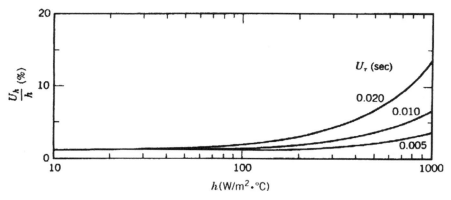

Figure 4.13   Uncertainty analysis results for the transient technique.

The results plotted in the figure show that the uncertainty in the experimental result increases with increasing h, which is opposite to the behavior seen for the steady-state technique. For the conditions and estimates used in this analysis, we can conclude that the transient technique will meet the requirements if the time constant can be measured with an uncertainty of 7 or 8 ms or less.

4-6.5 Implications of Uncertainty Analysis Results

The application of general uncertainty analysis to the two proposed techniques for determining h has given us information on the behavior of the uncertainty in h that is not immediately obvious from the data reduction equations [Eqs. (4.35) and (4.37)]. For the steady-state technique, the percentage uncertainty in h increases as h decreases (i.e., at lower air velocities). For the nominal values assumed, the uncertainty in the power measurement dominates for lower values of h, whereas the uncertainty in the differential temperature measurement dominates for higher values of h.

For the transient technique, the percentage uncertainty in h increases as h increases (i.e., at higher air velocities). For the nominal values assumed, the uncertainty in the measurement of the time constant dominates for higher h's, whereas the uncertainty in the value of the specific heat of the cylinder dominates for lower values of h.

Since the trends of the uncertainty in h differ for the two techniques, there is a range of h for which the steady-state technique gives "better" results and a range for which the transient technique gives better results. The crossover point at which the techniques give equal results from an uncertainty standpoint depends on the estimates of the uncertainties. Shown in Figure 4.14 is a comparison of the uncertainty analysis results for the two techniques with

$$U_W = 0.5 \text{ W} \qquad U_{\Delta T} = 0.25^\circ\text{C} \qquad U_\tau = 0.010 \text{ s}$$

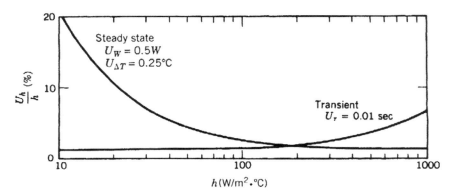

Figure 4.14 Comparison of uncertainty analysis results for the two proposed techniques.

and with the other uncertainties as estimated previously. For this case the transient technique is preferable for cases in which $h < 150$, the steady-state technique is preferable for cases in which $h > 200$, and the two techniques give results with about the same uncertainty for $150 < h < 200$ (where the units of h are $W/m^2 \cdot {}^{\circ}C$).

One qualification on these conclusions must be noted. In free (or natural) convection, h is a function of $T - T_a$ and is not constant during the cooling process of the transient technique. The assumptions used in obtaining the data reduction expression [Eq. (4.37)] for the transient technique are therefore violated in cases in which the heat transfer is by free convection. These cases occur when the forced-air velocity approaches zero.

4-7 EXAMPLES OF PRESENTATION OF RESULTS FROM ACTUAL APPLICATIONS

In this section, several different methods of presentation of the results of planning phase uncertainty analyses are shown from two research programs. The objective of the first program [7] was to determine the efficiency of a turbine tested in an airflow facility at temperatures near ambient. The objective of the second program [8] was to determine the specific impulse of a solar thermal absorber/thruster tested at a ground test facility.

4-7.1 Results from Analysis of a Turbine Test

The data reduction equation used to calculate turbine thermodynamic efficiency from measured test variables was derived from the basic definition of turbine efficiency: actual enthalpy change over ideal or isentropic enthalpy change. In this case the turbine was tested in an airflow facility at temperatures near ambient, so an ideal gas was assumed. The temperature drop across the turbine was measured to determine the actual enthalpy change ($\Delta h = C_p \, \Delta T$). Isentropic relations were used to write the ideal enthalpy change in terms of turbine inlet and exit total pressure rather than temperature. With these assumptions and substitutions, the equation for thermodynamic efficiency is

$$\eta_{th} = \frac{T_{01} - T_{02}}{T_{01}[1 - (P_{02}/P_{01})^{(\gamma-1)/\gamma}]} \tag{4.45}$$

where the subscript 0 indicates total conditions, 1 indicates the turbine inlet, and 2 indicates the turbine exit. The variable γ was assumed constant and to have negligible uncertainty in this analysis.

The general uncertainty analysis was used in a parametric study that considered six different turbine speeds and four different pressure ratios across the turbine, making a total of 24 test set points investigated in the analysis. The UMF results for each variable are shown in Figure 4.15, with set points 1 to 6 being at a

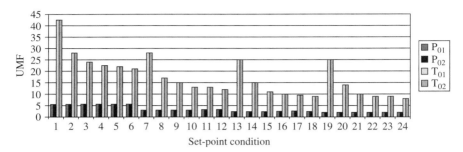

Figure 4.15 Presentation of UMF results for thermodynamic method of determining turbine efficiency for 24 different set points. (From Ref. 7.)

constant pressure ratio and arranged in order of increasing turbine speed, set points 7 to 12 at a second pressure ratio, and so on. Presented in this manner, the results of the analysis clearly show that the temperature measurements are potentially much more critical than the pressure measurements for accurately obtaining the turbine efficiency using the thermodynamic method. In this instance the UPC results confirmed this behavior, but that is not always the case, as we will see in Section 4-7.2.

The UPC results for each variable are shown in Figure 4.16, with the presentation of set points the same as in Figure 4.15. This figure clearly shows that for the particular set points and uncertainty estimates considered the accuracy of the temperature measurements is critical, with the exit temperature being the most important. It also shows that the accuracy of the exit pressure is slightly more significant than that of the inlet pressure. One could, of course, extend the analysis by investigating ranges of uncertainties for the measured variables.

4-7.2 Results from Analysis of a Solar Thermal Absorber/Thruster Test

In a solar thermal propulsion system, solar radiation is focused into an absorber/thruster whose walls are cooled by hydrogen flowing through tubes

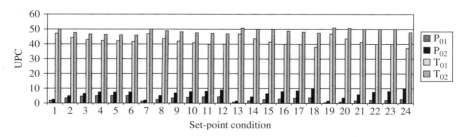

Figure 4.16 Presentation of UPC results for thermodynamic method of determining turbine efficiency for 24 different set points. (From Ref. 7.)

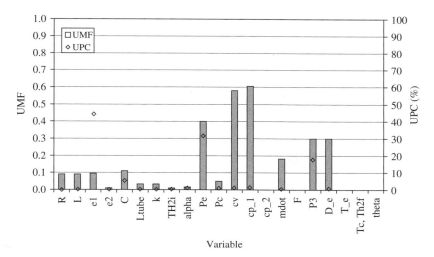

Figure 4.17   Presentation of UMF and UPC results for the specific impulse of a solar thermal absorber/thruster at one set point. (From Ref. 8; originally published by American Institute of Aeronautics and Astronautics.)

wound about its exterior. The hot hydrogen gas then enters a chamber and is exhausted through a nozzle, thus producing thrust. Such a system was to be tested in a ground test facility, and the specific impulse (thrust divided by weight flow rate of the propellant) was to be determined from measured variables. In the study cited, six potential data reduction equations were analyzed for two proposed test conditions.

Results for one data reduction equation at one set point are presented in Figure 4.17, where both the UMF and UPC values are presented for the measured variables, which are arranged along the abscissa. This type of presentation quickly allows one to see the implications of the uncertainty analysis results. In this case, the UMF analysis identified the uncertainties in specific heats (cv and cp_1) as potentially dominant. However, once uncertainty estimates were made for the variables and the UPC values were found, it was obvious that the specific heat uncertainties were of little concern and the uncertainty in the emissivity (e1) of the absorber cavity wall was the largest contributor to the uncertainty in specific impulse.

4-8 APPLICATION IN VALIDATION: ESTIMATING UNCERTAINTY IN SIMULATION RESULT DUE TO UNCERTAINTIES IN INPUTS

In Chapter 6, we will discuss uncertainty applications in the validation of a simulation result, specifically determination of the range for the modeling error in a simulation result. A key component in the validation uncertainty will be the uncertainties that arise from the input variables to the simulation and how they affect the uncertainty in the simulation result. The methodology for determining the simulation result uncertainty u_{input} associated with the input variables is described in Chapter 6, but a similar methodology has been used in the past for engineering planning and design studies. These applications are similar to the general uncertainty analysis principles described in this chapter.

In these studies, the uncertainties associated with the input variables are estimated and then TSM or MCM approaches are used to determine the uncertainty in the simulation result. Some applications of this methodology are given in Ref. 9–13. In the references the input uncertainties are analyzed based on their effect on the uncertainty of the result, similar to the UMF and UPC approaches presented in Section 4-2.

REFERENCES

1. Shortley, G., and Williams, D., *Elements of Physics*, 4th ed., Prentice-Hall, Upper Saddle River, NJ, 1965.

2. Hodkinson, J. R., "The Optical Measurement of Aerosols," in Davies, C. N., Ed., *Aerosol Science*, Academic Press, San Diego, CA, 1966, pp. 287–357.

3. Ariessohn, P. C., Eustis, R. H., and Self, S. A., "Measurements of the Size and Concentration of Ash Droplets in Coal-Fired MHD Plasmas," in *Proceedings of the 7th International Conference on MHD Electrical Power Generation*, Vol. II, 1980, pp. 807–814.

4. Holve, D., and Self, S. A., "Optical Measurements of Mean Particle Size in Coal-Fired MHD Flows," *Combustion and Flame*, Vol. 37, 1980, pp. 211–214.

5. Moffat, R. J., "Using Uncertainty Analysis in the Planning of an Experiment," *Journal of Fluids Engineering*, Vol. 107, 1985, pp. 173–178.

6. Incropera, F. P., and Dewitt, D. P., *Fundamentals of Heat Transfer*, Wiley, New York, 1981.

7. Hudson, S. T., and Coleman, H. W., "A Preliminary Assessment of Methods for Determining Turbine Efficiency," AIAA Paper 96–0101, American Institute of Aeronautics and Astronautics, New York, 1996.

8. Markopolous, P., Coleman, H. W., and Hawk, C. W., "Uncertainty Assessment of Performance Evaluation Methods for Solar Thermal Absorber/Thruster Testing," *Journal of Propulsion and Power*, Vol. 13, No. 4, 1997, pp. 552–559.

9. Boyack, B. E., et al., "Quantifying Reactor Safety Margins Part 1: An Overview of the Code Scaling, Applicability, and Uncertainty Evaluation Methodology," *Nuclear Science and Design*, Vol. 119, No. 1, 1990, pp. 1–16.

10. Wilson, G. E., et al., "Quantifying Reactor Safety Margins Part 2: Characterization of Important Contributions to Uncertainty," *Nuclear Science and Design*, Vol. 119, No. 1, 1990, pp. 17–32.

11. Wuff, W., et al., "Quantifying Reactor Safety Margins Part 3: Assessment and Ranging of Parameters," *Nuclear Science and Design*, Vol. 119, No. 1, 1990, pp. 33–65.

12. Taylor, R. P., Hodge, B. K., and Steele, W. G., "Series Piping System Design Program with Uncertainty Analysis," *Heating/Piping/Air Conditioning*, Vol. 65, No. 5, May 1993, pp. 87–93.

13. Taylor, R. P., Luck, R., Hodge, B. K., and Steele, W. G., "Uncertainty Analysis of Diffuse-Gray Radiation Enclosure Problems," *Journal of Thermophysics and Heat Transfer*, Vol. 9, No. 1, 1995, pp. 63–69.

PROBLEMS

4.1 Show by direct application of Eq. (4.6) to the data reduction equation

$$r = k(X_1)^a (X_2)^b (X_3)^c$$

that Eq. (4.19) is obtained. Consider a, b, c, and k to be constants that may be positive or negative.

4.2 The ideal gas equation of state can be written as

$$pV = mRT$$

For the nominal conditions and 95% confidence uncertainty estimates given below, what is the uncertainty in the volume of the gas?

$$p = 820 \pm 15 \text{ kPa}$$

$$m = 2.00 \pm 0.02 \text{ kg}$$

$$T = 305 \pm 3 \text{ K}$$

$$R = 0.287 \text{ kJ/kg} \cdot \text{K} \quad \text{(assume known perfectly)}$$

4.3 The velocity V of a body traveling in a circular path can be measured with an uncertainty estimated at 4%. The radius R of the circular path can be measured within 2%. What is the uncertainty associated with determination of the normal acceleration a_n from

$$a_n = \frac{V^2}{R}$$

Would it make much difference if the radius could be measured within 1%? within 0.5%?

4.4 A column of length L and square cross section ($b \times b$) is clamped at both ends. It has been proposed that the elastic modulus E of the column material be determined by axially loading the column until it buckles and using

$$P_{cr} = \frac{4\pi^2 EI}{L^2}$$

where P_{cr} is the buckling load and the moment of inertia I is given by

$$I = \tfrac{1}{12} b^4$$

How well can E be determined if b, L, and P_{cr} are measured to within 1%? What is the relative importance of the uncertainties in the measured variables?

4.5 A refrigeration unit extracts energy from a cold space at T_c and transfers energy as heat to a warm environment at T_h. A Carnot refrigerator defines the ideal case, and its coefficient of performance is defined by

$$\text{C.O.P.} = \frac{T_c}{T_h - T_c}$$

Assume equal uncertainties (in $°C$) in the measurement of T_c and T_h. How well must the temperature measurements be made for this ideal C.O.P. to be determined to within 1% when the nominal values of T_c and T_h are $-10°C$ and $20°C$?

4.6 A venturi flowmeter is to be used to measure the flow of air at low velocities, with air mass flow rate being given by

$$\dot{m} = CA\left[\frac{2p_1}{RT_1}(p_1 - p_2)\right]^{1/2}$$

where C is an empirically determined discharge coefficient, A is the throat flow area, p_1 and p_2 are the upstream and throat pressures, T_1 is the absolute upstream temperature, and R is the gas constant for air. We want to calibrate this meter and to determine C to within 2%. For the following nominal values and estimated uncertainties

$$p_1 = 30 \pm 0.5 \text{ psia}$$

$$T_1 = 70 \pm 3°F$$

$$\Delta p = p_1 - p_2 = 1.1 \pm 0.007 \text{ psi} \quad \text{(measured directly)}$$

$$A = 0.75 \pm 0.001 \text{ in.}^2$$

what is the allowable percentage uncertainty in the $\dot{m}$ measurement? That is, "how good" must the mass flowmeter used in the calibration process be?

4.7 For the same nominal conditions as in problem 4.6 the venturi flowmeter is used to determine the air mass flow rate. If the value of C is known $\pm1\%$, what will be the uncertainty in the determination of $\dot{m}$?

4.8 A heat transfer experiment is to be run with the anticipated behavior given by

$$\overline{T} = e^{-t/\tau}$$

where $\overline{T}$ is a nondimensional temperature that varies from 0 to 1 and can be measured $\pm2\%$, t is time in seconds and can be measured $\pm1\%$, and τ is the time constant that is to be determined from the experiment. Determine the behavior of the uncertainty in τ over the range of possible values for $\overline{T}$. Plot U_τ/τ versus $\overline{T}$ for $\overline{T}$ from 0 to 1.

4.9 For the radial conduction of heat through a pipe wall, the thermal resistance (per unit length) is given by

$$R_{\text{th}} = \frac{\ln(r_2/r_1)}{2\pi k}$$

where r_1 and r_2 are the radii of the inner and outer surfaces, respectively, of the pipe wall and k is the thermal conductivity of the pipe material. Perform a general uncertainty analysis and then a parametric study to find the sensitivity of the uncertainty in R_{th} to the radius ratio r_2/r_1. (You might want to initially assume 1% uncertainties for the values of r_1, r_2, and k.)

5

DETAILED UNCERTAINTY ANALYSIS: DESIGNING, DEBUGGING, AND EXECUTING AN EXPERIMENT

5-1 USING DETAILED UNCERTAINTY ANALYSIS

In Chapter 3 we presented the basic methodology involved in both general and detailed uncertainty analysis, and here it is appropriate to consider how these techniques are useful in different phases of an experimental program and how they fit together with the concepts of different orders of replication level that we discussed previously. An overview of some of the uses of uncertainty analysis in the different phases of an experimental program is given in Table 5.1.

As we saw in Chapter 4, general uncertainty analysis is used in the planning phase of an experimental program and is useful in ensuring that a given experiment can successfully answer the question of interest. Some decisions in the preliminary design of an experiment can be made based on the results of a general uncertainty analysis.

Once past the planning and preliminary design phases, the effects of systematic errors and random errors are considered separately using the techniques of detailed uncertainty analysis presented in this chapter. This means that estimates of systematic uncertainties and random uncertainties will be made and used in the design phase, then in the construction, debugging, execution, and data analysis phases, and finally in the reporting phase of an experiment, as shown in the table. If one thinks about it, it soon becomes obvious that there will almost always be much more information available with which to make the systematic and random uncertainty estimates in the later phases of an experiment than in the earlier phases. This means that estimates of a particular systematic uncertainty or random

Table 5.1   Uncertainty Analysis in Experimentation

Phase of Experiment	Type of Uncertainty Analysis	Uses of Uncertainty Analysis
Planning	General	Choose experiment to answer a question; preliminary design.
Design	Detailed	Choose instrumentation (zeroth-order estimates); detailed design (Nth-order estimates).
Construction	Detailed	Guide decisions on changes, etc.
Debugging	Detailed	Verify and qualify operation; first-order and Nth-order comparisons.
Execution	Detailed	Balance checks and monitoring operation of apparatus; choice of test points run.
Data analysis	Detailed	Guide to choice of analysis techniques.
Reporting	Detailed	Systematic uncertainties, random uncertainties, and overall uncertainties reported.

uncertainty may very well change as an experimental program progresses and we obtain more information—this should be expected and accepted as a fact. *Systematic uncertainties and random uncertainties must be estimated using the best information available at the time. Lack of information is no excuse for not doing an uncertainty analysis in the early phases of an experimental program—it is simply a reason why the estimates may not be as good as they will be later in the program.*

The manner in which these estimates are used can differ in timewise and sample-to-sample experiments. Recall that timewise experiments are those in which generally a given entity is tested, either as a function of time or at some steady-state condition in which data are taken over some period of time. Examples would be testing the performance characteristics of a given engine, determining the friction factor for a given pipe over a range of Reynolds numbers, and determining the heat transfer coefficient from a given object immersed in a fluid over a range of flow conditions. Sample-to-sample experiments are those in which generally some characteristic is determined for sample after sample, often with the variability from sample to sample being significant. In this case, sample identity can be viewed as analogous to time in a timewise experiment. Examples would be determining the heating value of a certain type of coal, determining the ultimate strength of a certain alloy, or determining some physical characteristic of a manufactured product for quality control purposes.

In the early stages of the design phase of a program, estimates of the systematic uncertainties and random uncertainties at the zeroth-order replication level are useful in choosing instrumentation and measurement systems. For timewise experiments, this means making the estimates while hypothesizing a totally steady process and environment. For sample-to-sample experiments, it means making the estimates assuming a single fixed sample. The zeroth-order systematic and random uncertainty estimates indicate the "best case" for a given measurement system in both types of experiment.

When we move beyond this stage in a timewise experiment, we make estimates at the first-order and Nth-order levels. Here we consider all the factors that will influence the systematic and random errors in the experiment. At the first-order replication level, we are interested in the variability of the experimental results for a given experimental apparatus. The descriptor for this variability is s_r, the random standard uncertainty of the result. In a timewise experiment, comparison of the estimated s_r and the random uncertainty determined from the scatter in multiple results at a given set point of the experimental apparatus is useful in the debugging phase. This is discussed in detail with an example in Section 5-6.

In a sample-to-sample experiment, first-order estimates of s_r made before multiple samples are tested are often not very useful, since the variation from sample to sample is usually unknown and is one of the things we discover with the experiment. After multiple samples have been tested, the difference (in a root-sum-square sense) between the calculated s_r and the zeroth-order random uncertainty estimate can be used as an estimate of the random uncertainty contribution due to sample-to-sample variability. This is illustrated in the lignite heating value determination example in Section 5-5.

In asking questions and making comparisons at the Nth-order replication level, we are interested in the interval within which "the truth" lies. This interval is described by U_r, the overall uncertainty in the result, which is found by combining the first-order random standard uncertainty s_r and the systematic standard uncertainty b_r using Eq. (3.19):

$$U_r = 2(b_r^2 + s_r^2)^{1/2}$$

for a large-sample 95% level of confidence.

The information available for estimating the uncertainties in a given phase of a test program varies depending on the type of test. In a test of a new model (somewhat similar to previous models) in a well-established wind tunnel facility, at each stage of the new test one might have excellent previous information on which to base estimates of random uncertainties of measured variables and results (e.g., drag coefficient) and estimates of systematic uncertainties. On the other hand, in setting up a completely new test with new equipment and control systems in a new facility one would expect initial estimates of the uncertainties to change significantly through the course of the test program as better information became available during calibrations, debugging, and so on.

In this chapter we discuss the application of detailed uncertainty analysis and illustrate its use with many examples. In Section 5-2, an overview of the complete (large-sample) methodology is presented. In Section 5-3, determination of the random uncertainty of a result is discussed and illustrated with three examples from real test programs that had different degrees of control—a venturi meter calibration facility, a laboratory-scale ambient temperature flow facility, and a rocket engine ground test facility. In Section 5-4, determination of the systematic uncertainty of a result is discussed, beginning with consideration of elemental error sources for a single variable, then situations in which different variables have correlated systematic errors, and finally comparative testing situations that can take advantage of such error correlations. In Section 5-5, a comprehensive example is presented using data from a real sample-to-sample experiment, and in Section 5-6 a comprehensive example is presented of the debugging and qualification of a real timewise experiment. Section 5-7 concludes with presentation of some additional considerations in experiment execution.

5-2 DETAILED UNCERTAINTY ANALYSIS: OVERVIEW OF COMPLETE METHODOLOGY

Now we turn our attention to detailed uncertainty analysis. Using this approach, the systematic and random errors in each measured variable are considered separately, and the resulting systematic and random uncertainties of the experimental result are investigated separately. The primary reason for considering the more complex approach of detailed uncertainty analysis is that it is very useful in the design, construction, debugging, data analysis, and reporting phases of an experiment to consider separately the effects of the errors that vary and those that do not—the random uncertainty and the systematic uncertainty of a result are estimated using different approaches. Although we use the word *complex* in connection with detailed uncertainty analysis, this should not be a cause for concern or a hesitancy to use the approach. As shown in the following, the application of detailed uncertainty analysis is accomplished through a series of logical steps that are themselves fairly straightforward.

The situation that we wish to analyze is illustrated in Figure 5.1, which shows a flow diagram of an experiment in which multiple measured variables are used to determine an experimental result. Each measurement system that is used to measure the value of an individual variable X_i is influenced by a number of elemental error sources (as we first discussed in Section 1-3.1; refer to Figure 1.4). The effects of these elemental errors are manifested as a systematic error and a random error in the measured value of the variable. These errors in the measured values then propagate through the data reduction equation and yield the systematic and random errors in the experimental result.

When a result r is calculated from several measured variables using a data reduction equation, a general functional formulation is (as discussed in Chapter 3)

$$r = r(X_1, X_2, \ldots, X_J) \tag{5.1}$$

where the X's are the measured variables. The uncertainties in the measured variables cause uncertainty in the result, and this is often modeled using a propagation equation based on a Taylor series expansion (the TSM approach).

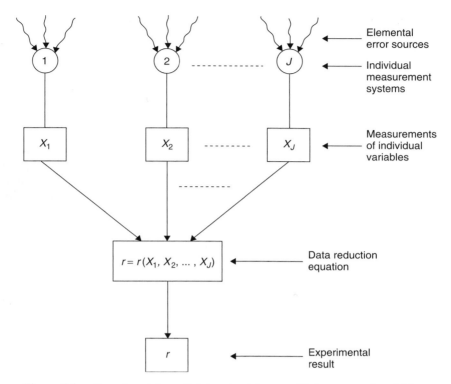

Figure 5.1 Experimental result determined from multiple measured variables.

The derivation and assumptions involved are discussed in Chapter 3 and in more detail in Appendix B.

Using the TSM approach, the systematic standard uncertainty b_r of the calculated result is given by

$$b_r^2 = \left(\frac{\partial r}{\partial X_1}\right)^2 b_{X_1}^2 + \left(\frac{\partial r}{\partial X_2}\right)^2 b_{X_2}^2 + \cdots + \left(\frac{\partial r}{\partial X_J}\right)^2 b_{X_J}^2 + \cdots$$
$$+ 2\left(\frac{\partial r}{\partial X_i}\right)\left(\frac{\partial r}{\partial X_k}\right) b_{X_i X_k} + \cdots \tag{5.2}$$

where the b_X's are the systematic standard uncertainties associated with the X variables and there is a correlated systematic error term containing a covariance factor $b_{X_i X_k}$ for each pair of variables that share common elemental systematic error sources. Correlated systematic errors occur often in engineering testing (such as when multiple transducers are calibrated with the same standard), and it is important to include consideration of the correlation terms in the b_r propagation equation. Evaluation of the correlated systematic error terms is discussed in detail in Section 5-4.2.

An analogous TSM equation for the random standard uncertainty s_r of the calculated result is

$$(s_r)^2_{\text{TSM}} = \left(\frac{\partial r}{\partial X_1}\right)^2 s^2_{X_1} + \left(\frac{\partial r}{\partial X_2}\right)^2 s^2_{X_2} + \cdots + \left(\frac{\partial r}{\partial X_J}\right)^2 s^2_{X_J} + \cdots$$
$$+ 2\left(\frac{\partial r}{\partial X_i}\right)\left(\frac{\partial r}{\partial X_k}\right) s_{X_i X_k} + \cdots \tag{5.3}$$

where there is a correlated random error term containing a covariance factor $s_{X_i X_k}$ for each pair of measured variables whose "random" variations are not independent of one another. These correlated random error terms have tradition-ally always been assumed to be zero, so that the propagation equation actually used is

$$(s_r)^2_{\text{TSM}} = \left(\frac{\partial r}{\partial X_1}\right)^2 s^2_{X_1} + \left(\frac{\partial r}{\partial X_2}\right)^2 s^2_{X_2} + \cdots + \left(\frac{\partial r}{\partial X_J}\right)^2 s^2_{X_J} \tag{5.4}$$

If the X values used in Eq. (5.1) are single measured values, the standard uncer-tainties in Eq. (5.4) are the individual variable standard deviation s_X values from Eq. (2.12):

$$s_X = \left[\frac{1}{N-1}\sum_{i=1}^{N}(X_i - \overline{X})^2\right]^{1/2}$$

where the N measured values of the variable X and thus s_X are known from previ-ous experiments. If the X values used in Eq. (5.1) are mean values $\overline{X}$ determined from N independent measurements of a variable X in the current experiment, the standard uncertainties in Eq. (5.4) are the $s_{\overline{X}}$ values from Eq. (2.14):

$$s_{\overline{X}} = \frac{s_X}{\sqrt{N}}$$

and the random uncertainty calculated from Eq. (5.4) is the random standard uncertainty of the mean result $(s_{\overline{r}})_{\text{TSM}}$.

As discussed briefly in Chapter 3, another way of estimating s_r in a steady-state test is by *direct calculation* using an equation analogous to Eq. (2.12):

$$(s_r)_{\text{direct}} = \left[\frac{1}{M-1}\sum_{i=1}^{M}(r_i - \overline{r})^2\right]^{1/2} \tag{5.5}$$

where there are M independent determinations of the result and

$$\overline{r} = \frac{1}{M}\sum_{i=1}^{M}r_i \tag{5.6}$$

An $(s_{\bar{r}})_{\text{direct}}$ can be defined using an equation analogous to Eq. (2.14),

$$s_{\bar{r}} = \frac{s_r}{\sqrt{M}} \tag{5.7}$$

As an example, suppose a result r is a function $r = r(y, z)$ of two measured variables y and z and both variables are measured at time $t = 1, 2, \ldots, 20$ s so that there are $M = 20$ measured data pairs. Values of s_y and s_z can be calculated from Eq. (2.12) and then $(s_r)_{\text{TSM}}$ determined using Eq. (5.4). A value of r can also be calculated at $t = 1, 2, \ldots, 20$ s so that one has a sample of M determinations of r. Then a direct calculation of s_r can be made using Eq. (5.5). Since the sample of M values of r implicitly contains the effects of any correlated variations among the measured variables, so does the $(s_r)_{\text{direct}}$ calculated using Eq. (5.5).

The direct approach has commonly not been used in single-test situations because unfortunately it has been presented in the past with nomenclature that caused a mind set that experimental results from *multiple tests* were necessary for its use. Clearly, the direct and the propagation approaches use exactly the same information and produce two estimates of the mean result and two estimates of the random standard uncertainty. It seems reasonable to calculate random standard uncertainty both ways in any steady-state test. In steady-state tests where the variations of the measured variables are not correlated, $(s_r)_{\text{TSM}}$ and $(s_r)_{\text{direct}}$ should be about equal (within the approximations inherent in the TSM equation). However, when the variations in different measured variables are not independent of one another, the values of s_r from the two equations can be significantly different. A significant difference between the two estimates may be taken as an indication of correlated variations among the measurements of the variables, and the direct estimate should be considered the more "correct" since it implicitly includes the correlation effects.

Examination of Eqs. (5.2) and (5.3) shows that each of the correlation terms contains a covariance factor ($b_{X_i X_k}$ or $s_{X_i X_k}$) multiplied by the product of the partial derivatives of the result with respect to the two variables. Assume the covariance factor is positive. If the partial derivatives are of opposite sign, the correlation term is negative and the effect of the correlation is to reduce the uncertainty. On the other hand, if the partial derivatives are of the same sign, the correlation term is positive and the effect of the correlation is to increase the uncertainty.

In Chapter 3, the TSM propagation approach was shown to produce expression (3.15) for the combined standard uncertainty of the result:

$$u_r = (b_r^2 + s_r^2)^{1/2}$$

where b_r is given by Eq. (5.2) and s_r is given by Eq. (5.4), (5.5), or (5.7) as appropriate. An expanded uncertainty which has an associated level of confidence was defined and details of its determination were discussed in Appendix B and in Chapter 3 for the general case in which the result r has an arbitrary number of degrees of freedom. From this point on, we will assume the number of degrees of freedom in the result is ≥ 9, that is, we make the "large-sample assumption," so that for a 95% level of confidence the expanded uncertainty of the result is given by Eq. (3.19):

$$U_r = 2\left(b_r^2 + s_r^2\right)^{1/2}$$

The MCM propagation approach is implemented as was shown in Figures 3.4 and 3.5. Once one has estimates of the elemental systematic uncertainties and assumed error distributions, it is very straightforward and (to the authors) easier to properly take into account the correlated systematic error effects than in the TSM.

In Section 5-3, we discuss determining the random uncertainty of a result and present several examples, including cases with correlated random errors. In Section 5-4, we discuss determining the systematic uncertainty of a result beginning with estimation of the elemental systematic uncertainties of a single variable and progressing to discussion of correlated systematic errors and their effects in single and in comparative experiments. We present in Section 5-5 a comprehensive example of the application of detailed uncertainty analysis in a real sample-to-sample experiment and in Section 5-6 a comprehensive example of its application in the debugging and qualification of a real timewise experiment. Section 5-7 concludes with some additional considerations applicable to experiment execution.

5-3 DETERMINING RANDOM UNCERTAINTY OF EXPERIMENTAL RESULT

In Section 2-3.1, where we first discussed estimating the standard deviation of a sample population, we mentioned the importance of the sample containing the influence of all of the variations of interest. In a sample-to-sample experiment, this means, for example, including the lot-to-lot variability if we want our estimated standard deviation to be representative of this effect—simply sampling many times within a given lot would not provide the standard deviation estimate appropriate for what was desired. In a timewise experiment, this means paying particular attention to the time scales of the things that cause variability. The situation is shown in Figure 5.2, which shows the variation of Y, which may be a variable or an experimental result, as a function of time.

Whether we are estimating s_x values for use in a propagation equation such as Eq. (5.4) or an s_r value using Eq. (5.5), the time period during which the sample is acquired is critical in determining exactly what s_x or s_r means. If, for instance, a sample data set is acquired over the time period Δt shown in the figure, neither the average value of Y nor the standard deviation calculated from the data set will be a good estimate of the values appropriate for a time period of 100 times Δt.

Data sets for determining estimates of standard deviations of measured variables or of experimental results should be acquired over a time period that is

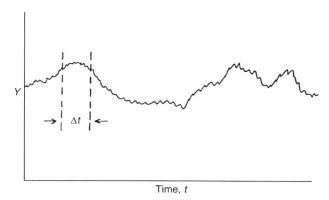

Figure 5.2 Variation with time of Y, a measured variable or an experimental result, for a "steady-state" condition.

large relative to the time scales of the factors that have a significant influence on the data and that contribute to the random uncertainty. In some cases, this means random uncertainty estimates must be determined from auxiliary tests that cover the appropriate time frame. This idea also comes into consideration when, in cases utilizing a test apparatus that has been used many times in the past, we may have previous data available with which to estimate the s_x or s_r values—the time period(s) over which the previous data were acquired determine what the standard deviation values really represent.

Barometric pressure varies with a time scale on the order of hours, as does relative humidity. If these factors are uncontrolled but do influence a variable in the experiment, this should be recognized. If an experiment is sensitive to the vibrations transmitted through the structure of a building, these vibrations may be quite different in morning and afternoon as various pieces of equipment in the building are operated or shut down. There are factors such as these that must be considered for each experiment in the proper determination of the random uncertainties for the measured variables and the experimental results.

There are also implications of these ideas in calculating and interpreting the $s_{\overline{X}}$ from Eq. (2.14) and $s_{\overline{r}}$ from Eq. (5.7). Using high-speed data acquisition systems, one could take an increasingly large number of measurements during the time period Δt shown in the Figure 5.2 and effectively make $s_{\overline{X}}$ and $s_{\overline{r}}$ approach zero. How useful these estimates would be from an engineering viewpoint would certainly be open to question.

As discussed in Section 5-1, the amount of information we have on which to base uncertainty estimates differs depending on the phase of the test program and the history of the test facility. In the initial phases of an experimental program before instrumentation is acquired and installed in a new test, we make 95% confidence estimates of random uncertainties in much the same manner as estimates of the systematic uncertainties are made. At this stage in the experiment, the measurement system may be the only random error source that we consider (as in a zeroth-order analysis). As a general rule of thumb, the 95% random uncertainty associated with the readability of an analog instrument can be taken as one-half of the least scale division. Of course, judgment must be used in applying this rule. For instance, if an analog instrument has a rather coarse scale division, the random uncertainty might be less than one-half of the least scale division. For a digital output, the minimum 95% random uncertainty in the reading resulting from the readability (assuming no flicker in the indicated digits) should be taken as one-half of the least digit in the output. Of course, the random uncertainty could be significantly less than this value. When there is no flicker in the output of a digital instrument (at a steady-state condition), the random errors are essentially damped by the digitizing process. Their magnitude is equal to or less than $\pm\frac{1}{2}$ the least digit. The significance of this potential random error should be considered in the early design phase of an experiment. If the instrument resolution is such that a random uncertainty estimate of $\pm\frac{1}{2}$ the least digit is unacceptably large, an instrument with a better resolution (more digits) should be considered for the experiment.

Another point related to the random uncertainty associated with instrument readability is its effect on the calibration of the measurement system. The systematic errors in the measurement system can be reduced by calibration. However, the minimum systematic uncertainty will be either that associated with the calibration standard or that from the instrument resolution, whichever is larger.

Later in the experimental program, when we are able to make measurements of a variable under conditions similar to those in the "real" experiment or to make multiple measurements of the variable during the actual experiment, the standard deviation s can be calculated from a data sample of N points. When such statistical determinations are made, it is very important to keep in mind the time period over which the N data points are acquired, as discussed previously.

In the following we present three examples of determining random uncertainties in real test programs that had different degrees of control—a compressible flow venturi meter calibration laboratory facility, a laboratory-scale ambient temperature flow facility housed in a building with modern heating and air conditioning, and a full-scale rocket engine ground test facility with the test stand exposed to the elements and the instrumentation and control systems housed in a bunker.

5-3.1 Example: Random Uncertainty Determination in Compressible Flow Venturi Meter Calibration Facility

A venturi [1] is being calibrated using an ASME nozzle as the flow standard, as shown in Figure 5.3. The venturi discharge coefficient determined in this test is a function of the standard flow rate W_{std}; the venturi inlet pressure and temperature, P_1 and T_1; the throat pressure P_2; and the inlet and throat diameters D_1 and D_2 as

$$C_d = f(W_{std}, P_1, P_2, T_1, D_1, D_2) \tag{5.8}$$

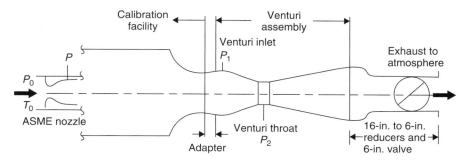

Figure 5.3   Venturi calibration schematic (not to scale). (From Ref. 1.)

In this example we consider only the random uncertainty in the discharge coefficient determination. This uncertainty is calculated using both the TSM propagation method [Eq. (5.4)] and the direct method [Eq. (5.5)].

The venturi was calibrated over a range of Mach numbers M and Reynolds numbers Re. Ten data scans were taken at each test condition. The venturi pressures and temperatures were recorded and used to calculate an average M, Re, and C_d for each test. The flow standard nozzle was operated at choked conditions for all test runs.

First, let us consider the TSM method for determining the random uncertainty for the discharge coefficient. It was assumed that only P_1, P_2, and T_1 contributed to the random uncertainty,

$$s_{C_d} = f(P_1, P_2, T_1)$$

since the random uncertainty associated with the mass flow rate of the standard was observed to be negligible relative to the random uncertainties of the test measurements of pressure and temperature. The means and standard deviations, s_i [Eq. (2.12)], of the three measurements (P_1, P_2, and T_1) were calculated from the 10 data scans for each test run ($N = 10$). The sensitivity coefficients θ_i were determined numerically using the expression for C_d where the values of P_1, P_2, and T_1 were set to the mean of the 10 scans for each test. The sensitivity coefficients and random uncertainties were then used in Eq. (5.4) to determine the random standard uncertainty for C_d in each test, as shown in Table 5.2.

The random uncertainty of the discharge coefficient was then determined directly from the C_d results. For each test C_d was calculated for each of the 10 data scans ($M = 10$), the standard deviation s_{C_d} was computed directly for the test using Eq. (5.5), and the mean of the ten C_d values was taken as the result for each test. These are also shown in Table 5.2.

The random uncertainties calculated by TSM propagation were larger than those calculated with the direct method. The differences were especially great at the low-Mach-number points. For low-Mach-number tests, the difference between

Table 5.2   **Random Uncertainty Values for C_d Using Both the TSM Propagation and the Direct Calculation Methods**

M	Re $\times$ 10^{-6}	C_d	$2s_{C_d}(\%)$ TSM	Direct	TSM/Direct
0.20	1.0	0.990	2.0	0.12	17
0.19	1.0	0.992	4.3	0.11	39
0.20	1.1	0.987	2.2	0.043	51
0.20	2.9	0.989	2.5	0.082	30
0.20	6.0	0.993	1.2	0.062	19
0.50	1.0	0.989	0.37	0.050	7
0.50	3.0	0.991	0.75	0.053	14
0.49	5.8	0.993	0.14	0.061	2
0.70	1.5	0.991	0.09	0.047	2
0.70	3.0	0.991	0.23	0.028	8
0.68	5.9	0.994	0.07	0.024	3

the venturi inlet and throat pressures was small. An examination of the sensitivity coefficients obtained from the propagation method revealed that the C_d random uncertainty was extremely sensitive to the venturi inlet and throat pressure measurements. The venturi temperature measurement had very little effect on the random uncertainty of C_d. Also, the venturi mass flow rate and, therefore, C_d are functions of the ratio of the inlet to throat pressures. This means that variations in the inlet pressure and throat pressure will not affect the random uncertainty of C_d if the ratio remains constant.

The propagation method treated the random uncertainties in the two venturi pressures as independent. A plot of these two pressures normalized to the critical flow nozzle inlet total pressure for a particular test (Figure 5.4) shows that the variations of the two pressures are not independent. This same trend was seen for all the test conditions. The fact that the throat pressure varied with the inlet pressure was a function of the test facility control. The distance between the ASME critical flow nozzle and the venturi was small; therefore, pressure variations in the choked flow and the venturi were in phase. The variations in the pressure measurements were not truly random; they were correlated. This correlation was not considered in the propagation method but was taken into account automatically in the direct method.

5-3.2 Example: Random Uncertainty Determination in Laboratory-Scale Ambient Temperature Flow Test Facility

In the laboratory-scale ambient temperature flow ejector experiment [2], compressed air was driven through a rocket nozzle embedded in a strut to produce a primary flow stream in an open-ended duct, as shown in Figure 5.5. The action of the air exhausting from the nozzle into the duct induces a secondary flow

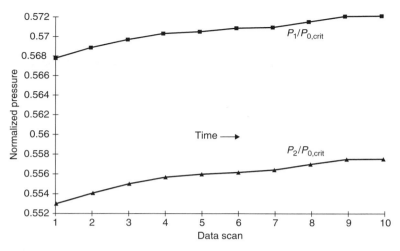

Figure 5.4   Normalized venturi inlet and throat pressures for a test. (From Ref. 1.)

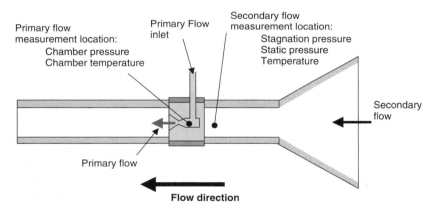

Figure 5.5   Diagram of ambient temperature flow ejector.

through the upstream open end of the duct. Measurements of the secondary-flow stagnation pressure and static pressure upstream of the strut, the inlet temperature of the secondary flow, and the rocket chamber pressure and temperature were made as indicated on the schematic. The experimental results discussed here are the mass flow rate of the induced secondary flow and the suction ratio.

The secondary mass flow rate $\dot{m}_s$ was determined assuming isentropic, incompressible, uniform flow of an ideal gas in the duct. The data reduction equation was

$$\dot{m}_s = P_1 A_{\text{duct}} \left(\frac{P_{01}}{P_1}\right)^{(\gamma-1)/(2\gamma)} \sqrt{\left(\frac{2\gamma}{RT_0(\gamma-1)}\right)\left(\left(\frac{P_{01}}{P_1}\right)^{(\gamma-1)/\gamma} - 1\right)} \quad (5.9)$$

where stagnation pressure P_{01} and static pressure P_1 were measured upstream of the strut, as indicated in Figure 5.5, and the stagnation temperature T_0 was measured at ambient conditions outside the duct.

The suction ratio for the ejector is defined as the ratio of the secondary mass flow rate to the primary mass flow rate,

$$\omega = \frac{\dot{m}_s}{\dot{m}_p} \tag{5.10}$$

and the primary mass flow rate $\dot{m}_p$ was determined from

$$\dot{m}_p = \frac{P_c}{T_c} A_t \sqrt{\frac{\gamma}{R}\left(\frac{2}{\gamma+1}\right)^{(\gamma+1)/(\gamma-1)}} \tag{5.11}$$

where P_c is the chamber pressure of the rocket nozzle, T_c is the chamber temperature, and A_t is the nozzle throat area. The suction ratio was thus determined from the data reduction equation

$$\omega = \frac{P_1 A_{\text{duct}}\left(\dfrac{P_{01}}{P_1}\right)^{(\gamma-1)/(2\gamma)} \sqrt{\left(\dfrac{2\gamma}{RT_0(\gamma-1)}\right)\left(\left(\dfrac{P_{01}}{P_1}\right)^{(\gamma-1)/\gamma} - 1\right)}}{\dfrac{P_c}{T_c} A_t \sqrt{\dfrac{\gamma}{R}\left(\dfrac{2}{\gamma+1}\right)^{(\gamma+1)/(\gamma-1)}}} \tag{5.12}$$

The independent, controlled variable in the experiment was the chamber pressure P_c, and the inevitable variations with time in this controlled pressure caused corresponding variations in P_{01} and P_1 upstream of the strut. The variations in the three pressures were therefore correlated, and this effect is not accounted for when Eq. (5.4) is used to calculate values of $(s_r)_{\text{TSM}}$.

Results from eight tests are considered here, with the set value of chamber pressure increasing with test number. The ratio of the values of random standard uncertainty calculated using the propagation and direct approaches is shown in Figure 5.6, where it is evident for these tests that the estimates of random standard uncertainty could be as much as 2 to $2\frac{1}{2}$ times too large if the traditional propagation approach is used instead of the direct approach of Eq. (5.5). The magnitude of this effect was not anticipated [2], as this experiment used state-of-the-art equipment and instrumentation and was in a well-controlled laboratory environment.

In the eight cold flow ejector tests, 2s intervals using $(s_r)_{\text{direct}}$ covered from 94 to 97% of secondary mass flow rate results and 95 to 97% of suction ratio results; equivalent coverage values for intervals using $(s_r)_{\text{TSM}}$ were 99 to 100% for secondary mass flow rate and 97 to 100% for suction ratio, which is consistent with the previous discussion that $(s_r)_{\text{TSM}}$ is erroneously too large in both cases.

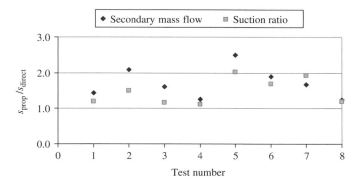

Figure 5.6   Ratios of random standard uncertainties from ejector tests.

5-3.3 Example: Random Uncertainty Determination in Full-Scale Rocket Engine Ground Test Facility

This example uses results from full-scale, hot-firing ground tests of a liquid rocket engine as reported in Ref. 2—such tests are obviously performed in a less well-controlled environment than are laboratory-scale tests. The average thrust and the specific impulse are the experimental results of interest here. In the tests the thrust was measured independently by two load cells, so the thrust was calculated as the average of the two measured thrust values F_1 and F_2 using the data reduction equation

$$F = 0.5(F_1 + F_2) \tag{5.13}$$

The specific impulse I was determined by dividing the thrust by the total weight flow rate of oxidizer and fuel using the data reduction equation

$$I = \frac{0.5(F_1 + F_2)}{(\rho_{ox} Q_{ox} + \rho_f Q_f)g} \tag{5.14}$$

where the subscript "ox" refers to oxidizer and "f" refers to fuel, ρ is density, Q is volumetric flow rate, and g is gravitational acceleration. The densities and g were taken from reference data and were considered as constants (with only systematic uncertainty) for a given test.

Shown in Figure 5.7 are measured values of thrust and volumetric flow rates during a "steady-state" period from one of the 17 rocket engine tests. In a test, the data were averaged over 1-s intervals and recorded throughout the test period. In the figure, the data have been normalized by dividing each recorded value by the test average, and they have also been shifted on the ordinate to more clearly show comparative trends.

The relative magnitudes of the random uncertainties determined using the TSM propagation method and the direct method are of interest, so $(s_r)_{TSM}$ and $(s_r)_{direct}$

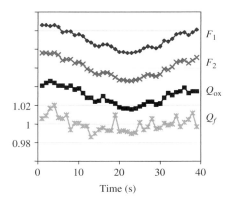

Figure 5.7   Normalized measurements from a liquid rocket engine ground test.

as given in Eqs. (5.4) and (5.5) are compared in the following analysis. Obviously, the ratio would remain the same if both were divided by $\sqrt{N}$ to produce $(s_{\bar{F}})_{\text{TSM}}$ and $(s_{\bar{F}})_{\text{direct}}$. Results from 17 tests of eight "identical" liquid rocket engines are considered. Data were recorded as 1-s averaged values (as shown in Figure 5.7), and the steady-state test period was taken as a 39-s interval after the initial start up transient.

For the thrust results for each test, s_F was determined using Eq. (5.4) and Eq. (5.5), and the results for the 17 tests are compared in Figure 5.8. In each test, $(s_F)_{\text{direct}}$ is greater than $(s_F)_{\text{prop}}$ by about 40%.

Consider the propagation equation (5.3), which contains the correlated random error term, applied to the data reduction equation for average thrust [Eq. (5.13)]:

$$(s_F)^2_{\text{TSM}} = \left(\frac{\partial F}{\partial F_1}\right)^2 s^2_{F_1} + \left(\frac{\partial F}{\partial F_2}\right)^2 s^2_{F_2} + 2\left(\frac{\partial F}{\partial F_1}\right)\left(\frac{\partial F}{\partial F_2}\right) s_{F_1 F_2} \qquad (5.15)$$

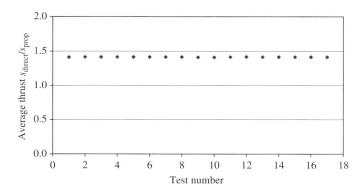

Figure 5.8   Ratios of random standard uncertainties for average thrust.

Both of the partial derivatives are equal to $+1/2$, and $s_{F_1 F_2}$ is positive since the outputs of the two load cells vary in unison. This means that the correlation term is positive, and when it is neglected as in Eq. (5.4) the random standard uncertainty that is calculated is erroneously low.

The random standard uncertainties calculated for specific impulse in the 17 tests are shown in Figure 5.9, and the behavior is the opposite of that observed for average thrust. For specific impulse, in each test $(s_I)_{TSM}$ is greater than $(s_I)_{direct}$. Considering $(s_I)_{direct}$ to be the "correct" estimate, the random standard uncertainty $(s_I)_{TSM}$ calculated using Eq. (5.4) is too large—by up to a factor of 4 in two of the tests. This means that the combination of the (neglected) correlation terms involving F_1, F_2, Q_{ox}, and Q_f is negative, leading to estimates that are too large when Eq. (5.4) is used.

When standard deviations are calculated from large samples, the interval $\bar{r} \pm 2s_r$ should contain roughly 95% of the sample of calculated results if the distribution of results is Gaussian. In the 17 rocket engine tests, intervals using $(s_r)_{direct}$ covered from 90 to 100% of average thrust results and 95 to 100% of specific impulse results; equivalent coverage values for intervals using $(s_r)_{TSM}$ were 82 to 95% for average thrust and 100% for specific impulse, which is consistent with the previous discussion that $(s_F)_{TSM}$ is erroneously too small and $(s_I)_{TSM}$ is erroneously too large.

5-3.4 Summary

The quality of the information on which random uncertainty estimates are based varies during the course of an experimental program, from what typically seems like insufficient information early in a new program in a new facility to the calculations based on the data obtained during the execution phase of the experiment. When one bases estimates of s_X and/or s_r on previous information from other

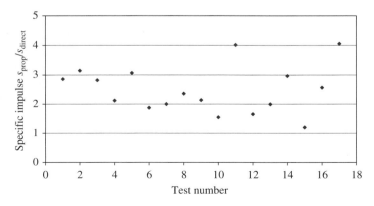

Figure 5.9   Ratios of random standard uncertainties for specific impulse.

experiments, it is important to recognize what sources of variability are represented in those estimates, keeping in mind the behavior illustrated in Figure 5.2.

The concept of time frame is also crucial when considering an averaged value $\overline{X}$ or $\overline{r}$ and the associated random uncertainties of the mean $s_{\overline{X}}$ and $s_{\overline{r}}$. In the authors' experience, in many timewise tests on complex engineering systems (such as one of the rocket engine tests in Section 5-3.3 or determination of the drag on a ship model during one run down a towing tank), the result of the test should be considered a *single* data point since it is acquired over a time period that is short compared to the variations that influence the system being tested. In such situations $s_{\overline{X}}$ and $s_{\overline{r}}$ have little engineering meaning or use, particularly if in the short period data are taken at a high rate and the $1/\sqrt{N}$ factor is used to drive $s_{\overline{X}}$ and $s_{\overline{r}}$ to arbitrarily small numbers.

During later phases of an experiment, when one can acquire meaningful current data, the random uncertainty of the result should be calculated both with a propagation approach such as Eq. (5.4) and the direct approach of Eq. (5.5). This gives insight into possible correlated time variations in multiple variables that are not obvious to a casual observer. The directly calculated value using Eq. (5.5) should be viewed as the "correct" value, keeping in mind the time scale considerations discussed above.

5-4 DETERMINING SYSTEMATIC UNCERTAINTY OF EXPERIMENTAL RESULT

In all of our discussions of systematic uncertainty we assume that corrections have been made for all of the systematic errors whose values are known, such as those determined by calibration. This means that the effects of all remaining significant systematic errors must be estimated. Because the "true" value of a measured variable is never known and because a systematic error does not vary at a given condition, there are no measurement or statistical procedures such as those of Chapter 2 that can be used generally to provide estimates of systematic uncertainties.

Shown schematically in Figure 5.10 are the steps in the procedure for determining b_r, the systematic uncertainty in the experimental result. Each measurement system used to determine the value of an individual variable is influenced by a number of elemental error sources. Systematic uncertainties are estimated for the elemental sources and combined to form the estimate of the systematic uncertainty for each measured variable. The systematic uncertainties of the individual variables are then propagated through the uncertainty analysis expression using the TSM or MCM approach to obtain the systematic uncertainty for the experimental result.

5-4.1 Systematic Uncertainty for Single Variable

The first step is to consider the elemental error sources that affect each of the measurements. These sources can be grouped into categories such as the following:

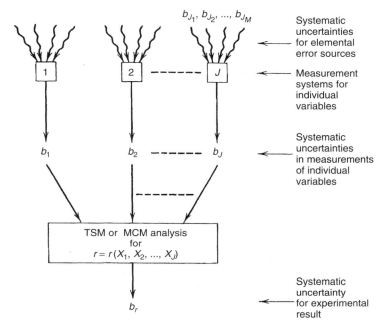

Figure 5.10 Detailed uncertainty analysis: determination of systematic uncertainty for experimental result.

1. *Calibration errors*. Some systematic error always remains as a result of calibration since no standard is perfect and no calibration process is perfect.
2. *Data acquisition errors*. There are potential errors due to environmental and installation effects on the transducer as well as in the system that acquires, conditions, and stores the output of the transducer.
3. *Data reduction errors*. Errors due to replacing calibration data with a curve fit, computational resolution, and so on.
4. *Conceptual errors*. Errors arising when the symbol in the data reduction equation is replaced by a measured value: for example, when a symbol representing an average is replaced with a point-measured value.
5. Others.

For instance, if for the Jth variable X_J there are M elemental systematic errors identified as significant and the corresponding systematic uncertainties are estimated as $b_{J_1}, b_{J_2}, \ldots, b_{J_M}$, the systematic uncertainty of the measurement of X_J is calculated as the root-sum-square (RSS) combination of the elemental systematic uncertainties:

$$b_J = \left[\sum_{k=1}^{M} b_{J_k}^2 \right]^{1/2} \tag{5.16}$$

The elemental systematic uncertainties b_{i_k} must be estimated for each variable X_i using the best information one has available at the time. In the design phase of an experimental program, manufacturers' specifications, analytical estimates, and previous experience will typically provide the basis for most of the estimates. As the experimental program progresses, equipment is assembled, and calibrations are conducted, these estimates can be updated using the additional information gained about the accuracy of the calibration standards, errors associated with the calibration process and curve-fit procedures, and perhaps analytical estimates of installation errors.

The calibration process should always be done with the measurement system (transducer, signal conditioner, data recording device, etc.) as close to the actual measurement condition and test installation arrangement as possible. The systematic error associated with the calibrated measurement system can then be reduced until it approaches that in the calibration standard, as discussed in Section 1-3.4.

In some cases, because of time or cost, the measurement system is not calibrated in its test configuration. In these cases the systematic errors that are inherent in the installation process must be included in the overall systematic uncertainty determination. Moffat [3] points out that these installation errors include interactions of the transducer with the system (such as radiation errors in temperature measurements) and system disturbances because of the presence of the transducer. (See asymmetric systematic uncertainties, Appendix E). These errors can sometimes be accounted for by modifying the data reduction equation to include these effects using a model. The systematic uncertainty estimates for the new terms in the modified data reduction equation then replace the estimates that would be made for the installation systematic errors.

In the absence of calibration, the instrument systematic uncertainty should be obtained from the manufacturer's information. This accuracy specification will usually consider such factors as gain, linearity, and zero errors, and it should be taken as a systematic uncertainty when no other information is available. This uncertainty will be in addition to installation uncertainties that may be present in the measurement.

Another point presented by Moffat is the idea of conceptual systematic errors. Are we really measuring the variable that is required in the data reduction equation? For instance, in fluid flow in a pipe, we might need the average velocity but only be able to measure the velocity at one point with the equipment available. The relationship between this single measurement of the velocity and the average velocity must be inferred from auxiliary information and included as a contributor to the systematic component in the uncertainty calculation.

With all of these possible sources of systematic error, one might decide that it is impossible to make accurate measurements. This is not the case. With experience in making specific measurements, one learns which errors are important and which errors are negligible. The key point is to think through the entire measurement process to account properly for the systematic errors.

5-4.1.1 Some Practical Considerations In the RSS method of combination, summing the squares of the elemental systematic uncertainties means that the larger elemental sources become even more dominant. A traditional "rule-of-thumb" is that any elemental uncertainty about one-fourth or less of the largest elemental uncertainty can be considered negligible. (This would not hold true, of course, if there were many of those with one-fourth the largest value. A rule-of-thumb must be considered a guide, not a law.) In the authors' experience, there are usually two to five significant elemental systematic uncertainties for a measured variable. *Resources should be used to obtain good uncertainty estimates for those elemental sources deemed significant based on initial order-of-magnitude estimates. Resources should not be wasted on obtaining good uncertainty estimates for insignificant sources—a practice that the authors have encountered too often, unfortunately.*

Example 5.1 The elemental systematic uncertainties (b_i) affecting measurements from a certain pressure transducer have been estimated as

$b_1 = 30$ Pa from the calibration standard

$b_2 = 50$ Pa from installation effects

$b_3 = 10$ Pa from replacing the calibration data with a curve fit

What is the appropriate estimate for the systematic standard uncertainty for a pressure measured with this transducer?

Solution From Eq. (5.16)

$$b_p = \sqrt{b_1^2 + b_2^2 + b_3^2}$$

or

$$b_p = \sqrt{30^2 + 50^2 + 10^2} = \sqrt{900 + 2500 + 100} = \sqrt{3500} = 59$$

What happens if the smallest source is neglected? Then

$$b_p = \sqrt{30^2 + 50^2} = \sqrt{900 + 2500} = \sqrt{3400} = 58$$

Considering how well we can estimate uncertainties, 58 and 59 are the same number.

So we see that any b_i whose value is one-fourth to one-fifth the value of the maximum source or less can be neglected. Of course, this rule of thumb would not be reasonable if there were 100 sources with $b = 10$! Engineering judgment must always be used.

In approaching the task of estimating a systematic uncertainty b, it is useful to remember that we are not trying to estimate the most probable value or the maximum value of the true but unknown systematic error β. Often, we try to estimate the limits of a Gaussian interval that contains the true β 95 times out of 100; that is, a $2b$ interval, as illustrated in Figure 5.11.

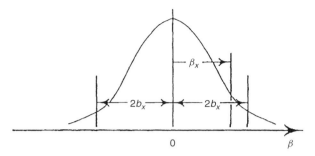

Figure 5.11 Systematic errors and systematic uncertainty for assumed Gaussian parent population for β.

Other distributions that are certainly reasonable to assume depending on the information available are

(1) a uniform (rectangular) distribution for β centered on zero with equal probability that β falls at any value between $-A$ and $+A$, which has a standard deviation (and thus value of b) equal to $A/\sqrt{3}$, or

(2) a triangular distribution symmetric about zero with base from $-A$ to $+A$, which has a standard deviation (and thus value of b) equal to $A/\sqrt{6}$.

Depending on the information available, sometimes we estimate b_X to be constant (i.e., not a function of X), and this value is often reported as a "% of full scale." Other times we estimate that b_X is a function of X—sometimes as a "% of reading" and sometimes as some other function of X. Any of these estimates must be viewed from a commonsense perspective. For instance, when b_X is specified as a "% of full scale" value, it is obvious that this estimate does not apply at or near $X = 0$, since if one nulls the instrument there should not be a systematic error of any significant magnitude at that point on the scale.

5-4.1.2 Digital Data Acquisition Errors

With an analog-to-digital (A/D) converter system, a calibrated voltage range is divided into discrete parts in which the resolution is dependent on the number of bits used by the unit. An 8-bit A/D converter will divide the voltage range into 2^8 parts, or 256, whereas a 12-bit system will improve the resolution to 2^{12}, or 4096, parts. The systematic error associated with the digitizing process will then depend on this resolution.

The 95% confidence uncertainty resulting from the digitizing process is usually taken to be $\frac{1}{2}$ LSB (least significant bit). The resolution of an LSB is equal to the full-scale voltage range divided by 2^N, where N is the number of bits used in the A/D converter. The systematic uncertainty associated with the digitizing process is then taken as $\frac{1}{2}$ LSB. This systematic uncertainty will be in addition to the other biases inherent in the instrumentation system (such as gain, zero, and nonlinearity errors).

Consider an example of an A/D converter system that has been calibrated for a range of 0 to 10 V full scale. With an 8-bit system, the 95% systematic uncertainty due to digitizing an input analog signal would be $\frac{1}{2}$ (10 V/256) or 0.020 V. For a 12-bit system the digitizing systematic uncertainty would be reduced to $\frac{1}{2}$ (10 V/4096) or 0.0012 V.

It should be stressed that these uncertainties are based on the full-scale calibration of the A/D converter. Care must be exercised in using the appropriately calibrated voltage range for the transducer being read. Consider, for example, a given transducer that has an output of 100 mV. If a 10-V A/D converter system were used to measure this signal, the digitizing systematic uncertainty for an 8-bit system would be 0.020 V or 20%. For a 12-bit system this uncertainty would be reduced to 1.2%. However, if a 100-mV calibration range were used for the A/D converter instead of 10 V, the digitizing systematic uncertainties would be very small for either an 8- or a 12-bit system. Both the voltage calibration range and number of bits used by the system must be considered when using digital data acquisition.

5-4.1.3 Property Value Uncertainty   In many experiments, values for some of the variables in the data reduction equations are not measured but rather are found from reference sources. This is often the case for material properties, which are typically tabulated as a function of temperature. Whether we enter the table to obtain a property value or use a curve-fit equation that represents the table, we obtain the same particular property value each time we use a given temperature. This value is not the true value—it is a "best estimate" based on experimental data and has an uncertainty associated with it.

As a specific example, consider that we need values for some thermal properties of air, say specific heat and thermal conductivity. Shown in Figures 5.12

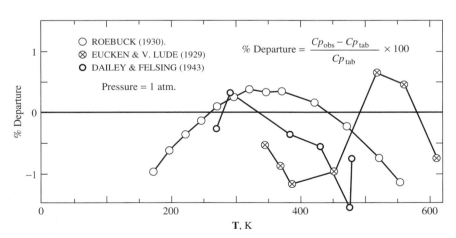

Figure 5.12   Departures of low-pressure experimental specific heats from tabulated values for air. (From Ref. 4.)

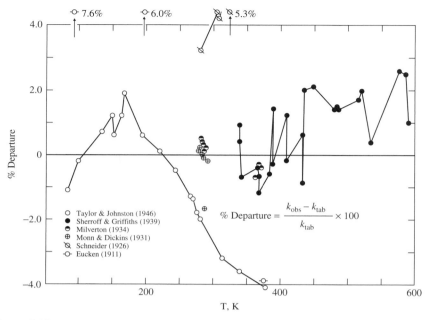

Figure 5.13 Departures of low-pressure experimental thermal conductivities from tabu-lated values for air. (From Ref. 4.)

and 5.13 are experimental data on specific heat and thermal conductivity, respec-tively, over a range of temperature. These are presented in the form of a percent-age departure of the experimental data from the values tabulated in the National Bureau of Standards (NBS) publication [4]. From these figures it should be obvious that assuming the uncertainties of property values are negligible is not generally a good assumption, even for such a common substance as air.

How should the random uncertainty and systematic uncertainty be estimated for a property value obtained from a table or curve-fit equation? Consider that once we choose a table or curve fit to use, we will always obtain *exactly* the same property value for a given temperature. If we entered the air specific heat table at 400 K a hundred times over the period of a month, each of the 100 specific heat values would be the same—there would be a zero random uncertainty associated with the value that we read from the table. This is the case regardless of the amount of scatter in the experimental data on which the table is based.

We conclude then that all of the errors (both random and systematic) in the experimental property data are "fossilized" [3] into a systematic error in values taken from the table or curve-fit equation that represents the experimental data. We thus should take our best estimate of the overall uncertainty in a property value and use that as the fossilized systematic uncertainty for that property. In practical terms, this generally means estimating a 95% uncertainty band based on the data scatter from different experiments, such as that seen in Figures 5.12 and 5.13.

5-4.2 Systematic Uncertainty of a Result Including Correlated Systematic Error Effects

Correlated systematic errors are those that are not independent of each other, typically a result of different measured variables sharing some identical elemental error sources. It is not unusual for the uncertainties in the results of experimental programs to be influenced by the effects of correlated systematic errors in the measurements of several of the variables. A typical example occurs when different variables are measured using the same transducer, such as multiple pressures sequentially ported to and measured with the same transducer or temperatures at different positions in a flow measured with a single probe that is traversed across the flow field. Obviously, the systematic errors in the variables measured with the same transducer are not independent of one another. Another common example occurs when different variables are measured using different transducers, all of which have been calibrated against the same standard, a situation typical of the electronically scanned pressure (ESP) measurement systems in wide use in aerospace test facilities. In such a case, at least a part of the systematic error arising from the calibration procedure will be the same for each transducer, and thus some of the elemental error contributions in the measurements of the variables will be correlated.

The b_{ik} terms in Eq. (5.2) must be approximated—there is in general no way to obtain the data with which to make a statistical estimate of the covariance of the systematic errors in X_i and in X_J. The approximation of such terms was considered in detail in Ref. 5, where it was shown that the approach that consistently gives the most satisfactory approximation for the correlated systematic uncertainties was

$$b_{ik} = \sum_{\alpha=1}^{L} (b_{i_\alpha})(b_{k_\alpha}) \tag{5.17}$$

where L is the number of elemental systematic error sources that are common for measurements of variables X_i and X_k.

If, for example,

$$r = r(X_1, X_2, X_3) \tag{5.18}$$

and it is possible for portions of the systematic uncertainties b_1, b_2, and b_3 to arise from the same sources, Eq. (5.2) gives

$$b_r^2 = \left(\frac{\partial r}{\partial X_1}\right)^2 b_1^2 + \left(\frac{\partial r}{\partial X_2}\right)^2 b_2^2 + \left(\frac{\partial r}{\partial X_3}\right)^2 b_3^2 \tag{5.19}$$

$$+ 2\left(\frac{\partial r}{\partial X_1}\right)\left(\frac{\partial r}{\partial X_2}\right) b_{12} + 2\left(\frac{\partial r}{\partial X_1}\right)\left(\frac{\partial r}{\partial X_3}\right) b_{13} + 2\left(\frac{\partial r}{\partial X_2}\right)\left(\frac{\partial r}{\partial X_3}\right) b_{23}$$

Note that there is one covariance term for each pair of variables whose measurements might share errors from identical sources.

For the sake of this discussion, assume that only measurements of X_1 and X_2 share some identical error sources; thus b_{13} and b_{23} will both be zero and Eq. (5.19) becomes

$$b_r^2 = \left(\frac{\partial r}{\partial X_1}\right)^2 b_1^2 + \left(\frac{\partial r}{\partial X_2}\right)^2 b_2^2 + \left(\frac{\partial r}{\partial X_3}\right)^2 b_3^2 + 2\left(\frac{\partial r}{\partial X_1}\right)\left(\frac{\partial r}{\partial X_2}\right)b_{12} \quad (5.20)$$

For a case in which the measurements of X_1 and X_2 are each influenced by four elemental error sources and sources 2 and 3 are the same for both X_1 and X_2, Eq. (5.16) gives

$$b_1^2 = b_{1_1}^2 + b_{1_2}^2 + b_{1_3}^2 + b_{1_4}^2 \quad (5.21)$$

and

$$b_2^2 = b_{2_1}^2 + b_{2_2}^2 + b_{2_3}^2 + b_{2_4}^2 \quad (5.22)$$

while Eq. (5.17) gives

$$b_{12} = b_{1_2}b_{2_2} + b_{1_3}b_{2_3} \quad (5.23)$$

If X_3 is influenced by three significant elemental error sources, Eq. (5.16) gives

$$b_3^2 = b_{3_1}^2 + b_{3_2}^2 + b_{3_3}^2 \quad (5.24)$$

and all the terms in Eq. (5.20) are known and b_r can be evaluated.

In the authors' experience, the only way to identify correlated systematic errors is to recognize elemental sources that are common for more than one measured variable and that therefore cause the measurements in the different variables to be either consistently high or consistently low. In such cases, the covariance estimator b_{ik} is always positive. In principle, it is possible for the true covariance of the systematic errors in different measured variables to be negative, but the authors have never encountered such a case (that they recognized).

Depending on the particular experimental approach, the effect of correlated systematic errors in the measurements of different variables can lead either to increased or to decreased systematic uncertainty in the final experimental result as compared to the same approach with no correlated systematic errors. Consider the final term in Eq. (5.20)—if some errors are correlated, the covariance b_{12} is nonzero (and positive). Thus, if the partial derivatives ($\partial r/\partial X_1$ and $\partial r/\partial X_2$) are of the same sign, the term is positive and b_r is increased. On the other hand, if the partial derivatives are of opposite signs, the term is negative and b_r is decreased. This observation suggests that the effect of correlated systematic errors can sometimes be used to advantage if the proper strategies are applied in planning and designing the experiment—sometimes one would want to force correlation of systematic errors using appropriate calibration approaches, sometimes not.

This point was evident in previously reported work [5] in which four experiments were investigated using a Monte Carlo simulation technique. In temperature difference, pressure coefficient, and compressor efficiency experiments, the presence of correlated systematic error effects decreased b_r, and it was shown that if the correlated effects were ignored, the resulting b_r estimate could be too large by several orders of magnitude. In the experiment in which the result was the average of three temperatures, the presence of correlated systematic error effects increased b_r, and it was shown that if the correlated effects were ignored, the resulting b_r estimate could be much too small.

The following examples illustrate the application of the ideas discussed previously.

5-4.2.1 Example: Correlated Errors in a Temperature Difference   Two plastic-encased thermistor probes are to be used to measure the inlet and outlet temperatures of water that flows through a heat exchanger (Figure 5.14). We wish to determine the systematic uncertainty in the temperature difference

$$\Delta T = T_2 - T_1$$

Solution. The probes are guaranteed by the manufacturer to match the "standard" resistance–temperature curve for this particular type of thermistor within $1.0°C$ over the temperature range of interest to us. We assume that this is a 95% confidence systematic uncertainty specification, so that averaging multiple readings would not reduce this number.

If this amount of systematic uncertainty is unsatisfactory for our purposes, we can calibrate the two probes by immersing them in a benchtop constant-temperature water or oil bath and using a calibrated thermometer (with $2b = 0.3°C$ systematic uncertainty) as a laboratory standard. The

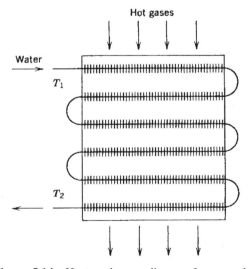

Figure 5.14   Heat exchanger diagram for example.

constant-temperature bath has maximum spatial temperature nonuniformity of $\pm 0.1°C$, according to the manufacturer, and we assume that this value is a reasonable 95% confidence systematic uncertainty estimate. (If we did not want to assume this, we could conduct a "side experiment" by moving the standard from point to point in the bath when it was at steady state. This would provide us with a sample distribution of local temperature differences and we could then calculate a standard deviation and use it as an estimate of b for the nonuniformity effect.)

After discussion with the heat exchanger manufacturer, we estimate (at 95% confidence) the elemental uncertainties due to water temperature nonuniformities at the inlet and outlet pipe cross sections where the probes will be placed as $0.1°C$ at the inlet and $0.2°C$ at the outlet. These account for the fact that the probes will probably not "see" the average water temperature at a cross section since the water temperature is not uniform across the cross section. Contributions from all other elemental error sources are considered to be negligible compared to those enumerated in the previous paragraphs.

The possible elemental sources for the two variables that will be measured in this experiment give the following:

	Systematic Uncertainty ($2b$)	
Elemental Source	$T_1(°C)$	$T_2(°C)$
1. As-delivered probe specification	1.0	1.0
2. Reference thermomenter—calibration	0.3	0.3
3. Bath nonuniformity—calibration	0.1	0.1
4. Spatial variation—conceptual	0.1	0.2

We can now use Eq. (5.16) and combine the uncertainties from the elemental sources 1 and 4 to obtain

$$b_{T_1} = [(0.5)^2 + (0.05)^2]^{1/2} = 0.5°C$$

$$b_{T_2} = [(0.5)^2 + (0.1)^2]^{1/2} = 0.5°C$$

as the systematic uncertainties for the two measured variables when we do not calibrate the probes. Similarly, we combine the uncertainties from the elemental sources 2, 3, and 4 to obtain

$$b_{T_1} = [(0.15)^2 + (0.05)^2 + (0.05)^2]^{1/2} = 0.17°C$$

$$b_{T_2} = [(0.15)^2 + (0.05)^2 + (0.1)^2]^{1/2} = 0.19°C$$

as the systematic uncertainties for the two measured variables when we do calibrate the probes.

The data reduction equation in this case is

$$\Delta T = T_2 - T_1$$

and the uncertainty analysis expression [from Eq. (5.2)] is

$$b^2_{\Delta T} = \left(\frac{\partial \Delta T}{\partial T_1}\right)^2 b^2_{T_1} + \left(\frac{\partial \Delta T}{\partial T_2}\right)^2 b^2_{T_2} + 2\left(\frac{\partial \Delta T}{\partial T_1}\right)\left(\frac{\partial \Delta T}{\partial T_2}\right)b_{T_1 T_2}$$

where

$$\frac{\partial \Delta T}{\partial T_1} = -1 \qquad \frac{\partial \Delta T}{\partial T_2} = 1$$

For the case in which we do not calibrate the probes, the errors affecting the probes are independent of one another and

$$b_{T_1 T_2} = 0$$

so that

$$b^2_{\Delta T} = (-1)^2 (0.5)^2 + (1)^2 (0.5)^2 \qquad b_{\Delta T} = 0.7°C$$

Now for the case in which we calibrate the probes, the uncertainty analysis expression becomes

$$b^2_{\Delta T} = (-1)^2 (0.17)^2 + (1)^2 (0.19)^2 + (2)(-1)(1)b_{T_1 T_2}$$

We will consider three possibilities:

1. If the probes were to be calibrated at different times against two different reference thermometers, then

$$b_{T_1 T_2} = 0$$

as there would be no correlation between any of the elemental systematic errors for T_1 and T_2 and therefore

$$b_{\Delta T} = 0.25°C$$

2. If the probes were to be calibrated at the same time against the same thermometer but were placed at different locations in the constant-temperature bath, the systematic error of the thermometer would affect each of the probes the same but that of the bath nonuniformity would not. In this case,

$$b_{T_1 T_2} = (0.15)(0.15) = 0.023$$

and we obtain

$$b^2_{\Delta T} = (-1)^2 (0.17)^2 + (1)^2 (0.19)^2 + (2)(-1)(1)(0.023)$$
$$= 0.0289 + 0.0361 - 0.046$$

or

$$b_{\Delta T} = 0.14°C$$

3. If the probes were to be calibrated at the same time against the same thermometer and were placed adjacent to each other at essentially the same location in the constant-temperature bath, the error of the thermometer and the error of the bath nonuniformity, sources 2 and 3, would affect each of the probes the same and would thus be correlated. In this case,

$$b_{T_1 T_2} = (0.15)(0.15) + (0.05)(0.05) = 0.025$$

so that
$$b_{\Delta T}^2 = (-1)^2(0.17)^2 + (1)^2(0.19)^2 + (2)(-1)(1)(0.025)$$
$$= 0.0289 + 0.0361 - 0.05$$

or
$$b_{\Delta T} = 0.12°C$$

This example illustrates the importance of considering the effect of correlated systematic errors. In this instance, proper design of the calibration procedure allowed the systematic uncertainty in the temperature difference ($0.12°C$) to be less than that ($0.17°C$ or $0.19°C$) in either of the measured temperatures.

5-4.2.2 Example: Correlated Errors in an Average Velocity   Consider the situation in which an average velocity of airflow at a duct cross section is needed so that a volumetric or mass flow rate can be determined. This occurs in a variety of heating, ventilating, and air conditioning (HVAC) applications, for example. In the determination of this average velocity of airflow in a duct, measurements are made at various positions in the duct cross section (either by traversing a probe or using a rake) and an area weighted-average velocity is determined as

$$V_{avg} = \frac{1}{A_{tot}} \sum_{i=1}^{N} A_i V_i$$

where each V_i is measured at the midpoint of the area A_i and A_{tot} is the duct cross-sectional area. If the values of all the A_i are equal, the average velocity becomes

$$V_{avg} = \frac{1}{N} \sum_{i=1}^{N} V_i$$

Applying Eq. (5.2), the systematic uncertainty for the average velocity due to the systematic uncertainties in the measurements of the velocities V_i is

$$b_{V_{avg}}^2 = \sum_{i=1}^{N} \left(\frac{1}{N}\right)^2 (b_{V_i})^2 + 2 \sum_{i=1}^{N-1} \sum_{k=i+1}^{N} \frac{1}{N}\frac{1}{N} b_{V_i V_k}$$

The values of b_{V_i} are all equal (to b_V, say) for cases where (1) a single probe is traversed (or a rake of probes is used with all transducers having been identically calibrated against the same standard), (2) the installation error is the same at each measurement position, and (3) the values of b_{V_i} are a fixed value ("% of full scale" rather than "% of reading"). For such cases Eq. (5.16) gives

$$b_{V_i V_k} = b_{V_i} b_{V_k} = b_V^2$$

and thus (leaving the algebra to the reader)

$$b_{V_{avg}} = b_V$$

For this set of circumstances, then, the systematic uncertainty for the average velocity is equal to the systematic uncertainty for a single measurement of velocity. Note, however, that this would not be the case if the values of b_{V_i} were specified as "% of reading" or some other function of the measured value.

On the other hand, if all the values of b_{V_i} are of the same magnitude (b_V) but are totally uncorrelated (perhaps when a rake of probes is used but none of the transducers have been calibrated against the same standard), the systematic uncertainty for the average velocity is

$$b_{V_{avg}} = \frac{1}{\sqrt{N}} b_V$$

There are several important points to be made from this example. First, the systematic uncertainties for V_{avg} for the two situations differ by a factor of $1/\sqrt{N}$, with the systematic uncertainty being smaller when there are no correlated systematic errors. *It cannot be said, therefore, that it is always "conservative" to ignore the correlated systematic error contributions.* Second, in HVAC applications minimum values of $N = 18$ for round ducts and $N = 25$ for rectangular ducts are recommended [6], meaning that ignoring the correlated systematic error effects would lead to estimates of $b_{V_{avg}}$ that were about one-fifth to one-fourth of the value estimated by proper consideration of the correlated systematic errors. This effect, then, can be quite considerable.

5-4.3 Comparative Testing and Correlated Systematic Error Effects[1]

In some experimental programs, the objective is to determine how the test result changes when modifications are made to the test article. Usually, for these situations, the test results are presented as a comparison to a baseline test. An

---

[1]This section is adapted from Ref. 7.

often-quoted misconception is that for such back-to-back tests using the same facility and instrumentation, the systematic uncertainty of the comparison is zero. Basically, these statements are an attempt to account for the correlated systematic errors in comparison tests without using the methodology discussed earlier. As shown below, statements that systematic errors cancel in comparative tests are correct only when specific conditions occur.

In this section we examine the effects of correlated systematic errors in two cases: (1) where the experimental result of interest is the difference of the results from two tests and (2) where the experimental result of interest is the ratio of the results from two tests. The effects can be most easily seen for the simple test in which the result is a function of two measured variables, so that the data reduction equation for a single test is given by

$$r = r(x, y) \tag{5.25}$$

5-4.3.1 Result Is a Difference of Test Results   Consider the case in which the experimental result of interest is the difference in the results of two tests. An example would be a wind tunnel test program to determine the drag increment due to a configuration change on a model for a given free-stream condition. Labeling the tests as a and b, the data reduction equation is

$$\delta = r(x_a, y_a) - r(x_b, y_b) \tag{5.26}$$

Application of (5.2) gives the expression for the systematic uncertainty for δ:

$$
\begin{aligned}
b_\delta^2 ={}& \left(\frac{\partial \delta}{\partial x_a}\right)^2 b_{x_a}^2 + \left(\frac{\partial \delta}{\partial x_b}\right)^2 b_{x_b}^2 + 2\left(\frac{\partial \delta}{\partial x_a}\right)\left(\frac{\partial \delta}{\partial x_b}\right) b_{x_a x_b} \\
&+ \left(\frac{\partial \delta}{\partial y_a}\right)^2 b_{y_a}^2 + \left(\frac{\partial \delta}{\partial y_b}\right)^2 b_{y_b}^2 + 2\left(\frac{\partial \delta}{\partial y_a}\right)\left(\frac{\partial \delta}{\partial y_b}\right) b_{y_a y_b} \\
&+ 2\left(\frac{\partial \delta}{\partial x_a}\right)\left(\frac{\partial \delta}{\partial y_a}\right) b_{x_a y_a} + 2\left(\frac{\partial \delta}{\partial x_b}\right)\left(\frac{\partial \delta}{\partial y_b}\right) b_{x_b y_b} \\
&+ 2\left(\frac{\partial \delta}{\partial x_b}\right)\left(\frac{\partial \delta}{\partial y_a}\right) b_{x_b y_a} + 2\left(\frac{\partial \delta}{\partial x_a}\right)\left(\frac{\partial \delta}{\partial y_b}\right) b_{x_a y_b}
\end{aligned} \tag{5.27}
$$

Derivatives with respect to a variable with an a subscript are evaluated at the conditions of test a, while derivatives with respect to a variable with a b subscript are evaluated at the conditions of test b. Noting that

$$
\begin{aligned}
\frac{\partial \delta}{\partial x_a} &= \frac{\partial r}{\partial x_a} & \frac{\partial \delta}{\partial y_a} &= \frac{\partial r}{\partial y_a} \\
\frac{\partial \delta}{\partial x_b} &= -\frac{\partial r}{\partial x_b} & \frac{\partial \delta}{\partial y_b} &= -\frac{\partial r}{\partial y_b}
\end{aligned} \tag{5.28}
$$

and substituting into Eq. (5.27) give

$$
b_\delta^2 = \left(\frac{\partial r}{\partial x_a}\right)^2 b_{x_a}^2 + \left(\frac{\partial r}{\partial x_b}\right)^2 b_{x_b}^2 + 2\left(\frac{\partial r}{\partial x_a}\right)\left(-\frac{\partial r}{\partial x_b}\right) b_{x_a x_b}
$$
$$
+ \left(\frac{\partial r}{\partial y_a}\right)^2 b_{y_a}^2 + \left(\frac{\partial r}{\partial y_b}\right)^2 b_{y_b}^2 + 2\left(\frac{\partial r}{\partial y_a}\right)\left(-\frac{\partial r}{\partial y_b}\right) b_{y_a y_b}
$$
$$
+ 2\left(\frac{\partial r}{\partial x_a}\right)\left(\frac{\partial r}{\partial y_a}\right) b_{x_a y_a} + 2\left(-\frac{\partial r}{\partial x_b}\right)\left(-\frac{\partial r}{\partial y_b}\right) b_{x_b y_b}
$$
$$
+ 2\left(-\frac{\partial r}{\partial x_b}\right)\left(\frac{\partial r}{\partial y_a}\right) b_{x_b y_a} + 2\left(\frac{\partial r}{\partial x_a}\right)\left(-\frac{\partial r}{\partial y_b}\right) b_{x_a y_b} \qquad (5.29)
$$

The third term on the right-hand side (RHS) accounts for the correlated systematic errors in the measurement of x in test a and the measurement of x in test b, and the sixth term does the same for the y measurements. Since in most comparative tests a variable will be measured by the same system in tests a and b, most (if not all) of the elemental error sources will be identical in the measurements of a variable in the two tests and the covariance estimators $b_{x_a x_b}$ and $b_{y_a y_b}$ will be nonzero.

The final four terms on the RHS account for the possibility that the x and y measurements might share some elemental error sources. This usually occurs when x and y are both pressures, or both temperatures, and so on, and are measured either with the same instrument or with different instruments that have been calibrated against the same standard. For this discussion, assume that measurements of x and measurements of y share no elemental error sources, so that the final four RHS terms are zero. Equation (5.29) then becomes

$$
b_\delta^2 = \left(\frac{\partial r}{\partial x_a}\right)^2 b_{x_a}^2 + \left(\frac{\partial r}{\partial x_b}\right)^2 b_{x_b}^2 - 2\left(\frac{\partial r}{\partial x_a}\right)\left(\frac{\partial r}{\partial x_b}\right) b_{x_a y_b}
$$
$$
+ \left(\frac{\partial r}{\partial y_a}\right)^2 b_{y_a}^2 + \left(\frac{\partial r}{\partial y_b}\right)^2 b_{y_b}^2 - 2\left(\frac{\partial r}{\partial y_a}\right)\left(\frac{\partial r}{\partial y_b}\right) b_{y_a y_b} \qquad (5.30)
$$

What are the conditions necessary for b_δ to be zero? First, all the elemental systematic error sources for the x measurements must be the same in tests a and b, and the systematic uncertainties for all the elemental sources must be constants (% of full-scale type) rather than a function of the measured value (% of reading type, for instance). The same must be true for the systematic errors in the y measurements. Then

$$
b_{x_a} = b_{x_b} = b_x \qquad b_{x_a x_b} = b_x^2
$$
$$
b_{y_a} = b_{y_b} = b_y \qquad b_{y_a y_b} = b_y^2
$$
$$\qquad (5.31)$$

Second, the derivatives evaluated at test a conditions must equal the corresponding derivatives evaluated at test b conditions; that is,

$$\frac{\partial r}{\partial x_a} = \frac{\partial r}{\partial x_b} \qquad \frac{\partial r}{\partial y_a} = \frac{\partial r}{\partial y_b} \tag{5.32}$$

For the conditions represented by Eqs. (5.31) and (5.32), the first, second, and third terms on the RHS of Eq. (5.30) add to zero, as do the fourth, fifth, and sixth terms. Equation (5.30) then yields $b_\delta = 0$.

Now consider the experiment mentioned previously investigating the drag increment due to a model configuration change. The data reduction equation is

$$\delta = \frac{F_a}{\frac{1}{2}\rho_a V_a^2 A_a} - \frac{F_b}{\frac{1}{2}\rho_b V_b^2 A_b} \tag{5.33}$$

The derivatives with respect to V, for example, are

$$\frac{\partial \delta}{\partial V_a} = \frac{-2F_a}{\frac{1}{2}\rho_a V_a^3 A_a} \qquad \frac{\partial \delta}{\partial V_b} = \frac{2F_b}{\frac{1}{2}\rho_b V_b^3 A_b} \tag{5.34}$$

Even if the denominators of the two derivatives are equal (tests a and b free-stream conditions *exactly* the same), the numerators will not be the same except for the very special instance of the dependent test variable (drag force in this case) having the same value in the two tests. This is a highly unlikely occurrence—even if a test is replicated with everything the same, including configuration, the measured values of the variables will be slightly different in tests a and b. In the more likely case of interest in this section, the drag forces F_a and F_b will be different, so the condition of Eq. (5.32) is not fulfilled and $b_\delta \neq 0$.

Based on the discussion above, it must be concluded that the effects of systematic uncertainties *do not* cancel out in comparative experiments in which the result is the difference in the results of two tests. When the same test apparatus and instrumentation are used in the two tests, however, the systematic uncertainty in δ [from Eq. (5.29) or (5.30)] can be significantly less than the systematic uncertainty in the result of either test a or b.

When considering the question, "How small an increment can be distinguished in a comparative test?" the *overall uncertainty* associated with δ must be considered, since it can be argued that the minimum distinguishable increment must certainly be larger than its uncertainty. From Eq. (3.15) the combined standard uncertainty in δ is

$$u_\delta^2 = b_\delta^2 + s_\delta^2 \tag{5.35}$$

where b_δ is given by Eq. (5.29) or (5.30). The determination of the appropriate estimate for s_δ warrants further discussion.

The estimate of s_δ must include the influence of *all* the significant factors that cause variation in δ, considering the process by which δ is determined. Consider a case similar to the drag increment experiment, for instance, in which the wind tunnel was shut down after test a; the model was removed, altered, and reinstalled; the tunnel was restarted and reset to the free-stream conditions of test a; and then the test b data were taken. For a DRE given by Eq. (5.26), using Eq. (5.4) to determine s_δ by propagation gives

$$s_\delta^2 = \left(\frac{\partial r}{\partial x_a}\right)^2 s_{x_a}^2 + \left(\frac{\partial r}{\partial x_b}\right)^2 s_{x_b}^2 + \left(\frac{\partial r}{\partial y_a}\right)^2 s_{y_a}^2 + \left(\frac{\partial r}{\partial y_b}\right)^2 s_{y_b}^2 \qquad (5.36)$$

This will not, in general, give an appropriate estimate of s_δ if the s_x and s_y values are estimated based on data taken within a test without tunnel conditions reset and model reinstallation effects included. A more appropriate estimate would be given by substituting

$$r_a = r(x_a, y_a) \qquad r_b = r(x_b, y_b) \qquad (5.37)$$

into Eq. (5.26) and applying Eq. (5.4) to yield

$$s_\delta^2 = (1)^2 s_{r_a}^2 + (-1)^2 s_{r_b}^2 \qquad (5.38)$$

where the s_r values are determined directly from a sample of multiple results that includes all the effects discussed above.

In many practical engineering tests (such as the drag increment experiment example), one might determine an estimate s_r directly from a sample of previous results (for a similar model if possible) that includes the effects of resetting tunnel conditions and model reinstallation and assume that

$$s_{r_a} = s_{r_b} = s_r \qquad (5.39)$$

so that Eq. (5.38) gives

$$s_\delta = \sqrt{2} s_r \qquad (5.40)$$

It should be noted that, depending on how the comparative experiments are performed, some factors listed above that can affect δ may not occur during the testing period. For instance, if the model angle is the only parameter changed between tests a and b with the wind tunnel continually operating at the same nominal conditions during both tests, the random uncertainty associated with the comparison would not contain the effects of shutting the tunnel down and restarting it or of model removal and reinstallation. The s_r values in Eq. (5.39) should represent only those random factors that exhibit variations in the comparison of results from tests a and b.

A special case of a difference-type comparative test is the comparison of two test results, either from the same facility or from different facilities, to determine if they represent the same phenomenon. In this case the difference, or the value of δ, is not the desired test result, but instead, a check is needed to see if the two tests have a reasonable likelihood of representing the same event. We can use the preceding discussion to consider this question. Basically, we can say that for large sample sizes, if the difference in the two results is δ [Eq. 5.26] and if $|\delta| < k u_\delta$ where u_δ is from Eq. (5.35), there is no indication that the two tests represent different physical phenomena. When the two tests are in the same facility, the correlated systematic error terms in Eq. (5.29) can be important. For tests in different facilities on different test articles, there probably will be no correlated systematic errors.

5-4.3.2 Result Is a Ratio of Test Results
Consider the case in which the experimental result of interest is the ratio of the results of two tests. Such cases occur, for example, in heat transfer testing when data are presented as ratios of heat transfer coefficients for different wall boundary conditions on a given test article. An example, including discussion of the uncertainty aspects, was published by Chakroun et al. [8]. Labeling the tests as a and b, the data reduction equation is

$$\eta = \frac{r(x_a, y_a)}{r(x_b, y_b)} \tag{5.41}$$

Application of Eq. (5.2) gives the expression for the systematic uncertainty for η:

$$
\begin{aligned}
b_\eta^2 =& \left(\frac{\partial \eta}{\partial x_a}\right)^2 b_{x_a}^2 + \left(\frac{\partial \eta}{\partial x_b}\right)^2 b_{x_b}^2 + 2\left(\frac{\partial \eta}{\partial x_a}\right)\left(\frac{\partial \eta}{\partial x_b}\right) b_{x_a x_b} \\
&+ \left(\frac{\partial \eta}{\partial y_a}\right)^2 b_{y_a}^2 + \left(\frac{\partial \eta}{\partial y_b}\right)^2 b_{y_b}^2 + 2\left(\frac{\partial \eta}{\partial y_a}\right)\left(\frac{\partial \eta}{\partial y_b}\right) b_{y_a y_b} \\
&+ 2\left(\frac{\partial \eta}{\partial x_a}\right)\left(\frac{\partial \eta}{\partial y_a}\right) b_{x_a y_a} + 2\left(\frac{\partial \eta}{\partial x_b}\right)\left(\frac{\partial \eta}{\partial y_b}\right) b_{x_b y_b} \\
&+ 2\left(\frac{\partial \eta}{\partial x_b}\right)\left(\frac{\partial \eta}{\partial y_a}\right) b_{x_b y_a} + 2\left(\frac{\partial \eta}{\partial x_a}\right)\left(\frac{\partial \eta}{\partial y_b}\right) b_{x_a y_b}
\end{aligned} \tag{5.42}
$$

Noting that

$$
\begin{aligned}
\frac{\partial \eta}{\partial x_a} &= \frac{1}{r_b}\frac{\partial r}{\partial x_a} & \frac{\partial \eta}{\partial y_a} &= \frac{1}{r_b}\frac{\partial r}{\partial y_a} \\
\frac{\partial \eta}{\partial x_b} &= \frac{-r_a}{r_b^2}\frac{\partial r}{\partial x_b} & \frac{\partial \eta}{\partial y_b} &= \frac{-r_a}{r_b^2}\frac{\partial r}{\partial y_b}
\end{aligned} \tag{5.43}
$$

and substituting into Eq. (5.42) give

$$
\begin{aligned}
b_\eta^2 =& \frac{1}{r_b^2}\left(\frac{\partial r}{\partial x_a}\right)^2 b_{x_a}^2 + \frac{r_a^2}{r_b^4}\left(\frac{\partial r}{\partial x_b}\right)^2 b_{x_b}^2 - 2\frac{r_a}{r_b^3}\left(\frac{\partial r}{\partial x_a}\right)\left(\frac{\partial r}{\partial x_b}\right) b_{x_a x_b} \\
&+ \frac{1}{r_b^2}\left(\frac{\partial r}{\partial y_a}\right)^2 b_{y_a}^2 + \frac{r_a^2}{r_b^4}\left(\frac{\partial r}{\partial y_b}\right)^2 b_{y_b}^2 - 2\frac{r_a}{r_b^3}\left(\frac{\partial r}{\partial y_a}\right)\left(\frac{\partial r}{\partial y_b}\right) b_{y_a y_b}
\end{aligned}
$$

$$+ 2\frac{1}{r_b^2}\left(\frac{\partial r}{\partial x_a}\right)\left(\frac{\partial r}{\partial y_a}\right)b_{x_a y_a} + 2\frac{r_a^2}{r_b^4}\left(\frac{\partial r}{\partial x_b}\right)\left(\frac{\partial r}{\partial y_b}\right)b_{x_b y_b}$$

$$- 2\frac{r_a}{r_b^3}\left(\frac{\partial r}{\partial x_b}\right)\left(\frac{\partial r}{\partial y_a}\right)b_{x_b y_a} - 2\frac{r_a}{r_b^3}\left(\frac{\partial r}{\partial x_a}\right)\left(\frac{\partial r}{\partial y_b}\right)b_{x_a y_b} \qquad (5.44)$$

This expression is identically zero only for the relatively uninteresting case in which tests a and b are exactly the same. In that case, $\eta = 1$ and the first, second, and third terms on the RHS of Eq. (5.44) add to zero, as do the fourth, fifth, and sixth terms and the seventh, eighth, ninth, and tenth terms. For a more practical situation, the study by Chakroun et al. [8] showed that when heat transfer test results were presented as a ratio to results from a baseline test, the systematic uncertainty of the ratio was less than the systematic uncertainty associated with the results from either test. The study included the complexity of the measurements of x and y variables sharing an identical source of systematic error, so that all six covariance terms in Eq. (5.44) were nonzero.

As in the case in which the experimental result was the difference in two results, when the same test apparatus and instrumentation are used in two tests, the ratio of the two test results can have a lower systematic uncertainty than the systematic uncertainty in either of the individual test results. The same random uncertainty considerations discussed for the difference tests apply in the case of ratio tests.

5-5 COMPREHENSIVE EXAMPLE: SAMPLE-TO-SAMPLE EXPERIMENT

A mining company has retained our laboratory to determine the heating value of some lignite in which they have an interest. Lignite is a brownish coal that is relatively high in ash and moisture content. The most abundant supplies of lignite in the United States are in the Gulf Coast region and in the Northern Great Plains region. The company wants to develop a North Dakota lignite deposit into a boiler fuel.

5-5.1 Problem

Initially, the company wants to determine the heating value of a small portion of the deposit of lignite. They plan to deliver sealed samples to our laboratory from this section of the deposit. From these they want answers to the following questions:

1. How well can we measure the heating value of a single sample?
2. What is the mean heating value for the particular portion of the deposit, and what is the uncertainty of this mean value?

In question 1 we are essentially considering the uncertainty at a zeroth-order replication level as we are concerned only with the uncertainty caused by the systematic and random errors in the instrumentation system. To answer question 1 we consider the systematic and random uncertainties in the measurement system that combine to give the uncertainty in a single heating value determination.

We know that lignite can have significant variations in the ash and moisture content even within a single deposit. For this reason we expect that there will be variations in the heating value determinations among the samples because of the physical variability of the material. Therefore, the uncertainty found for question 1 will probably be different from that found for question 2. The uncertainty found for question 2 will essentially be at an Nth-order replication level at which variations in the material are taken into account along with the systematic and random uncertainties in the measurement system.

5-5.2 Measurement System

The heating value will be determined with an oxygen bomb calorimeter (see Figure 5.15). This device is a standard, commercially available system for making this type of determination. A measured sample of the fuel is enclosed in the metal pressure vessel that is then charged with oxygen. This bomb is then placed in a bucket containing 2000 g of water and the fuel is ignited by means of an electrical fuse. The heat release from the burning fuel is absorbed by the metal bomb and

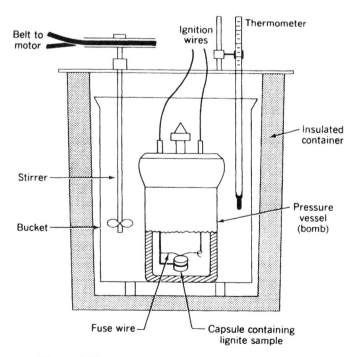

Figure 5.15 Schematic of oxygen bomb calorimeter.

the water. By measuring the temperature rise $(T_2 - T_1)$ of the water, the heating value of the fuel can be determined from an energy balance on the calorimeter, which gives the expression

$$H = \frac{(T_2 - T_1)C - e}{M} \qquad (5.45)$$

where H is the heating value (cal/g), T is temperature (°F or °R), C is the calibrated energy equivalent of the calorimeter (cal/°R), e is the energy release from the burned fuse wire (cal), and M is the mass of the fuel sample (g). Note that this is the higher heating value as the vapor formed is condensed to liquid.

In the discussion that follows, we first consider the application of detailed uncertainty analysis at the zeroth-order replication level. Next we consider the phase of the experiment in which heating values have been determined from multiple samples. These data are then used to calculate the random uncertainty associated with the measurement system and with the ash and moisture variations in the samples—this corresponds to analysis at the first-order replication level. Finally, we look at the Nth-order replication level at which the random uncertainties from the variability of the material and from the measurement system and the systematic uncertainties from the measurement system are combined to give us the uncertainty in the mean heating value for the portion of the lignite deposit being analyzed.

5-5.3 Zeroth-Order Replication-Level Analysis

The heating value of each lignite sample will be determined from data obtained with the bomb calorimeter system using the data reduction equation (5.45):

$$H = \frac{(T_2 - T_1)C - e}{M}$$

The calibration "constant" for the system was determined previously by using a high-purity benzoic acid as the fuel. In this calibration, multiple tests were run and the mean value for C was found to be 1385 cal/°R. The random standard uncertainty of this mean value was ± 4 cal/°R, and the systematic standard uncertainty for the calibration constant was determined to be ± 1 cal/°R. Therefore, the combined standard uncertainty in the calibration constant is

$$u_c = [(4)^2 + (1)^2]^{1/2} = 4.1 \text{ cal/°R} \qquad (5.46)$$

When C is used in Eq. (5.45), the uncertainty associated with that number will always be 4.1 cal/°R. Even though this uncertainty was originally obtained from combining random and systematic uncertainties, it becomes "fossilized" into a systematic standard uncertainty $b_{C,A}$, for all future calculations when the fixed value of $C = 1385$ cal/°R is used.

Note that we have used the systematic standard uncertainty symbol, b_C, for the calibration constant uncertainty but that we have added a comma A to denote that the source of the uncertainty is classified as ISO type A (calculated from data). This double notation can be useful for providing both the engineering aspects of the uncertainty and the ISO evaluation method.

When heating value measurements are made with the bomb calorimeter, the system is identical to that used for calibration. Therefore, the systematic uncertainties in the temperatures, the mass, and the fuse wire energy release have been replaced by the uncertainty in the calibration constant.

The temperatures T_1 and T_2 are measured with a thermometer that has $0.05°F$ as its least scale division. To determine the random uncertainty for measuring temperatures with this thermometer, different people were asked to read a steady temperature. The random standard uncertainty calculated from these readings was $0.026°F$. Therefore, a single temperature measurement will have this value, $s_{T,A}$, as the estimate of its random standard uncertainty.

The fuse wire comes with a scale marked in calories with a least scale division of 1.0 cal. With the scale division, we estimate the 95% random uncertainty for the measurement of a length of wire to be 0.5 cal and the random standard uncertainty to be one-half of that value, or 0.25 cal. The fuse wire correction e used in Eq. (5.45) is the difference between the original wire length and the length of the unburned wire recovered from the bomb after the test is run. Using Eq. (5.4), the random standard uncertainty for e, $s_{e,B}$, is then the square root of the sum of the squares of the random standard uncertainties for the two wire lengths, or 0.35 cal. The subscript B is included because the random uncertainty for the fuse wire energy release is estimated rather than being calculated from data.

The mass is measured with a digital balance that has a resolution of 0.0001 g. From experience in observing the "jitter" in the digital readout, we estimate that the 95% random uncertainty for a mass measurement is 0.0002 g, yielding a random standard uncertainty of 0.0001 g. The mass M in Eq. (5.45) is similar to the fuse wire length in that it is the difference between two measurements: the mass of the container with the lignite sample in it and the mass of the empty sample container. Therefore, the random standard uncertainty for M, $s_{M,B}$, is 0.00014 g [using Eq. (5.4) to combine the two mass random uncertainties]. The zeroth-order estimates of the systematic and random standard uncertainties for the variables are summarized in Table 5.3.

Table 5.3 Zeroth-Order Estimates of Systematic and Random Standard Uncertainties for Variables in Heating Value Determination

Variable	Systematic Standard Uncertainty	(ISO Type)	Random Standard Uncertainty	(ISO Type)
T_1	—	—	$0.026°F$	A
T_2	—	—	$0.026°F$	A
C	4.1 cal/°R	A	—	—
e	—	—	0.35 cal	B
M	—	—	0.00014 g	B

Referring back to the heating value expression (5.45),

$$H = \frac{(T_2 - T_1)C - e}{M}$$

the partial derivatives needed for the uncertainty analysis are

$$\frac{\partial H}{\partial T_2} = \frac{C}{M} \tag{5.47}$$

$$\frac{\partial H}{\partial T_1} = -\frac{C}{M} \tag{5.48}$$

$$\frac{\partial H}{\partial C} = \frac{T_2 - T_1}{M} \tag{5.49}$$

$$\frac{\partial H}{\partial e} = -\frac{1}{M} \tag{5.50}$$

$$\frac{\partial H}{\partial M} = \frac{-[(T_2 - T_1)C - e]}{M^2} = -\frac{H}{M} \tag{5.51}$$

The uncertainty analysis expression that must be used for determining b_H is then given by Eq. (5.2), so that

$$b_H^2 = \left(\frac{\partial H}{\partial C} b_C \right)^2 \tag{5.52}$$

Substituting for the partial derivative we get

$$b_H^2 = \left(\frac{T_2 - T_1}{M} \right)^2 b_C^2 \tag{5.53}$$

We can see from Eq. (5.53) that nominal values of the test variables must be found before a numerical value for b_H can be determined. In this instance we have available the results from a previous test, so we use those values (Table 5.4) along

Table 5.4   Nominal Values from Previous Test

Measured Variable	Value
T_1	75.25°F
T_2	78.55°F
C	1385 cal/°R
e	12 cal
M	1.0043 g

$$H = \frac{(T_2 - T_1)C - e}{M} = 4539 \text{ cal/g}$$

with the systematic standard uncertainty estimate in Table 5.3. The systematic uncertainty in the result then becomes

$$b_H^2 = 726 \tag{5.54}$$

or

$$b_H = 13 \text{ cal/g} \tag{5.55}$$

From this calculation, we see that the "fossilized" uncertainty in the calibration constant determines the systematic uncertainty in the heating value.

The random standard uncertainty of the result is found using Eq. (5.4), which gives

$$s_H^2 = \left(\frac{\partial H}{\partial T_2} s_{T_2}\right)^2 + \left(\frac{\partial H}{\partial T_1} s_{T_1}\right)^2 + \left(\frac{\partial H}{\partial C} s_C\right)^2$$
$$+ \left(\frac{\partial H}{\partial e} s_e\right)^2 + \left(\frac{\partial H}{\partial M} s_M\right)^2 \tag{5.56}$$

Substituting for the partial derivatives yields

$$s_H^2 = \left(\frac{C}{M}\right)^2 s_{T_2}^2 + \left(\frac{C}{M}\right)^2 s_{T_1}^2 + \left(\frac{T_2 - T_1}{M}\right)^2 s_C^2$$
$$+ \left(\frac{1}{M}\right)^2 s_e^2 + \left(\frac{H}{M}\right)^2 s_M^2 \tag{5.57}$$

Using the values given in Tables 5.3 and 5.4, the random uncertainty becomes

$$\begin{array}{ccccc} (T_2) & (T_1) & (C) & (e) & (M) \end{array}$$
$$s_H^2 = 1286 + 1286 + 0 + 0.12 + 0.4 \tag{5.58}$$

or

$$s_H = 51 \text{ cal/g} \tag{5.59}$$

Here the temperature random uncertainty is controlling.

The original question of how well we can measure the heating value of a single sample can now be answered. The uncertainty (with a 95% confidence) in the heating value determination is found using Eq. (3.19) as

$$U_H = 2(b_H^2 + s_H^2)^{1/2} \tag{5.60}$$

Using the systematic and random standard uncertainties calculated previously, the uncertainty becomes

$$U_H = 2[(13)^2 + (51)^2]^{1/2} = 105 \text{ cal/g} \tag{5.61}$$

This uncertainty, which represents the zeroth-order replication-level estimate for this experiment, is about 2% of the lignite heating value.

5-5.4 First-Order Replication-Level Analysis

After the mining company delivered 26 sealed samples to the laboratory, the heating value of each was determined using the bomb calorimeter. These heating value results are summarized in Table 5.5.

Because we have multiple results, we can use the techniques of Section 5-3 and directly calculate a random uncertainty and mean value for H. Using Eq. (5.6), the mean of these heating value results is

$$\overline{H} = 4486 \text{ cal/g} \tag{5.62}$$

From Eq. (5.5), the standard deviation of the collection of 26 results is

$$s_H = 87.2 \text{ cal/g} \tag{5.63}$$

This random standard uncertainty is greater than that the 51 cal/g found for a single sample, which seems reasonable as it includes contributions to the random

Table 5.5   Heating Value Results for 26 Lignite Samples

Sample Number	Heating Value (cal/g)
1	4572
2	4568
3	4547
4	4309
5	4383
6	4354
7	4533
8	4528
9	4383
10	4546
11	4539
12	4501
13	4462
14	4381
15	4642
16	4481
17	4528
18	4547
19	4541
20	4612
21	4358
22	4386
23	4446
24	4539
25	4478
26	4470

uncertainty from both the measurement system and the material composition variation from sample to sample. It is the random uncertainty appropriate for use in an analysis at the first-order replication level.

If we divide the contributors to the first-order random standard uncertainty into two categories—zeroth-order and sample-to-sample material variation—we can write

$$(s_H)_{1st}^2 = (s_H)_{zeroth}^2 + (s_H)_{mat}^2 \tag{5.64}$$

Since the first-order random uncertainty has been determined as 87.2 cal/g and the zeroth-order random uncertainty has been determined as 51 cal/g, the random standard uncertainty due to sample-to-sample variation in the lignite composition can be found as

$$
\begin{aligned}
(s_H)_{mat} &= [(s_H)_{1st}^2 - (s_H)_{zeroth}^2]^{1/2} \\
&= [(87.2)^2 - (51)^2]^{1/2} = 71 \text{ cal/g}
\end{aligned}
\tag{5.65}
$$

This estimate is useful because it tells us that the random standard uncertainty due to the variability of lignite composition within the portion of the deposit investigated is actually greater than the random uncertainty associated with our measurement system.

The analysis presented in this section illustrates the primary use of an analysis at the first-order level in a sample-to-sample experiment. We were able to estimate the effect of the material variations from sample to sample once we had a zeroth-order random uncertainty estimate and a first-order random uncertainty calculated directly from multiple results.

5-5.5 *N*th-Order Replication-Level Analysis

Now we are in a position to answer the second of the original two questions. The mining company wants us to determine the mean heating value of the lignite in the portion of the deposit being analyzed along with the uncertainty of the mean. With the 26 samples that were tested, we found the mean of the results to be 4486 cal/g.

The uncertainty in this mean result is, from Eq. (3.19),

$$U_{\overline{H}} = 2(b_{\overline{H}}^2 + s_{\overline{H}}^2)^{1/2} \tag{5.66}$$

where

$$s_{\overline{H}} = \frac{s_H}{\sqrt{M}} \tag{5.67}$$

Substituting for s_H from Eq. (5.63) we obtain

$$s_{\overline{H}} = \frac{87.2 \text{ cal/g}}{\sqrt{26}} = 17.1 \text{ cal/g} \tag{5.68}$$

and

$$U_{\overline{H}} = 2[(13)^2 + (17.1)^2]^{1/2} = 43 \text{ cal/g} \qquad (5.69)$$

where the systematic standard uncertainty found at the zeroth-order replication level in Eq. (5.55) applies here as well. Therefore, the true mean lignite heating value is, with a 95% level of confidence, in the range 4486 ± 43 cal/g, or about 4500 cal/g, with an uncertainty of about 1%.

It should be noted that in this experiment the only significant systematic error source was present at the zeroth-order replication level. This made the Nth-order systematic uncertainty the same as the zeroth-order systematic uncertainty. This is not true in general, as often systematic errors arise due to installation, sensor/environment interaction, and other sources not considered at the zeroth level. In general, $(b_r)_{N\text{th}} > (b_r)_{\text{zeroth}}$.

5-6 COMPREHENSIVE EXAMPLE: DEBUGGING AND QUALIFICATION OF A TIMEWISE EXPERIMENT

After proceeding through the planning and design phases of an experiment, choosing the instrumentation, and building up the experimental apparatus, typically a debugging and qualification phase is necessary. In this phase we try to determine and fix the unanticipated problems that occur and reach the point at which we feel confident that the experiment is performing as anticipated. In this section we discuss making checks at the first-order level of replication to investigate the run-to-run scatter in timewise experiments. If our first-order random uncertainty estimate s_r accounts for all the significant sources of random error, the random standard uncertainty determined from the run-to-run scatter in the debugging phase of the experiment should agree with s_r. We also discuss using the Nth-order expanded uncertainty estimate (U) to compare results against accepted results or theoretical values. This can sometimes be done only for some limiting case or condition that covers a portion of the range of operation of the experiment, but nonetheless it serves as a valuable qualification check for the experimental effort.

5-6.1 Basic Ideas

The idea of different orders of replication level is very useful in the debugging and qualification stages of a timewise experiment. The way we have defined the first-order level of replication, its primary use is in checking that we understand and have considered the effects of the repeatability or scatter in the results r_i of a timewise experiment.

The utility of a first-order replication comparison in the debugging phase of an experiment uses the following logic. If all the factors that influence the random error of the measured variables and the result have been accounted for properly in determining s_r, then the random standard uncertainty obtained from the scatter in the results at a given set point should be approximated by s_r. Here s_r is the random uncertainty of the result determined from propagating the estimated random uncertainty of the measured variables using the TSM or an s_r directly calculated previously in similar tests. If the random uncertainty in the results is greater than the anticipated values, this indicates that there are factors influencing the random error of the experiment that are not properly accounted for. This should be taken as an indication that additional debugging of the experimental apparatus and/or experimental procedures is needed.

The usefulness of the Nth-order replication level in the debugging/qualification of an experiment is in comparing current experimental results with theoretical results or previously reported, well-accepted results from other experimental programs. Since comparisons at the Nth-order level are viewed as comparisons of intervals within which "the truth" lies with 95% confidence, agreement should be expected within the overall uncertainty interval ($\pm U_r$) of the result.

Agreement at the Nth-order level can be taken as an indication that all significant contributors to the uncertainty in the experimental result have been accounted for. Often this comparison, and hence the conclusion, can be made over only a portion of the operating domain of the experiment.

5-6.2 Example

As an example of an experimental program [9] that used checks at both the first- and Nth-order replication levels to advantage, consider the wind tunnel system shown in Figure 5.16. This system was designed to investigate the effect of surface roughness on the heat transfer between a turbulent boundary layer airflow

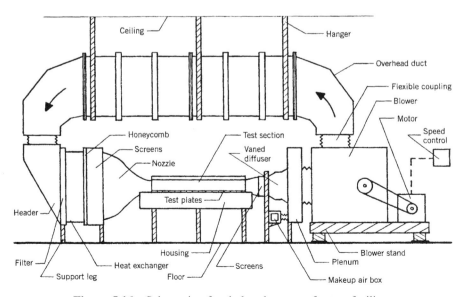

Figure 5.16   Schematic of turbulent heat transfer test facility.

and a flat plate, which in this case was the bottom surface of the wind tunnel test section. The basic experimental approach was the steady-state technique described in Section 4-6. Twenty-four plate segments made up the test surface, each with its own electrical heater/power supply/instrumentation system. Each plate segment (Figure 5.17) was supplied the power W necessary to maintain its temperature at a constant prescribed value T_w. The convective heat transfer coefficient h was determined in nondimensional form as a *Stanton number*, defined by

$$St = \frac{h}{\rho_\infty u_\infty c_{p\infty}} \tag{5.70}$$

where ρ_∞ is the air density, $c_{p\infty}$ the air constant-pressure specific heat, and u_∞ the free-stream air velocity.

Upon using the defining equation for h,

$$q_{\text{convection}} = hA(T_w - T_\infty) \tag{5.71}$$

and an energy balance on a test plate segment that included conduction losses q_c and radiation losses q_r, the data reduction equation became

$$St = \frac{W - q_r - q_c}{\rho_\infty u_\infty c_{p\infty} A(T_w - T_\infty)} \tag{5.72}$$

A test consisted of setting the desired value of free-stream velocity u_∞ and waiting until the computer-controlled data acquisition and control system adjusted the power to each plate so that all 24 plates were at a steady-state temperature T_w. Once such a steady state was reached, values of all variables were recorded and the data were reduced and plotted in log-log coordinates as St versus Re_x,

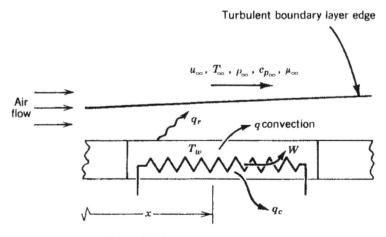

Figure 5.17 Test plate schematic.

where the Reynolds number is given by the second data reduction equation in the experiment,

$$\mathrm{Re}_x = \frac{\rho_\infty u_\infty x}{\mu_\infty} \tag{5.73}$$

and x is the distance from the leading edge of the test surface to the midpoint of a particular plate. The initial qualification testing was planned to use a smooth test surface and five values of u_∞ from 12 to 67 m/s.

A detailed uncertainty analysis during the design phase using the techniques outlined in this chapter concluded that random uncertainties in the determination of the Stanton number were negligible compared to the systematic uncertainties, so that

$$s_{\mathrm{St}} \approx 0 \tag{5.74}$$

and

$$(2b)_{\mathrm{St}} \approx U_{\mathrm{St}} \approx 2 \text{ to } 4\% \tag{5.75}$$

depending on the experimental set point.

This uncertainty analysis result indicated, then, that replications on different days at a particular set point (u_∞) should produce St data with negligible variation. That is, a check at the first-order replication level during the debugging phase should show negligible run-to-run scatter. Figure 5.18 shows replicated data

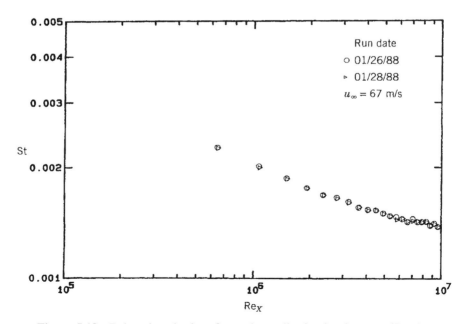

Figure 5.18   Debugging check at first-order replication level, $u_\infty = 67$ m/s.

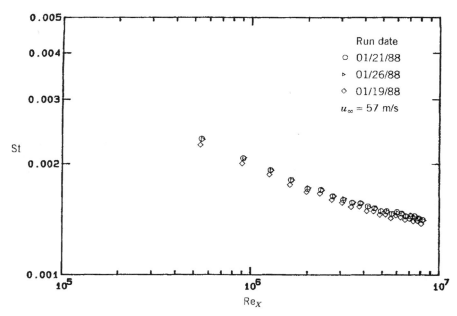

Figure 5.19 Debugging check at first-order replication level, $u_\infty = 57$ m/s. Problem indicated since $s_{St} \approx 0$ was expected.

at $u_\infty = 67$ m/s. These results support the conclusion of the detailed uncertainty analysis that $s_{St} \approx 0$. Figure 5.19, however, presents three replications at the $u_\infty = 57$ m/s set point that indicate a problem. The St data for the 01/19/88 run are obviously different from the runs made on 01/21/88 and 01/26/88—the scatter for St indicated by these three replications is definitely not zero. It therefore could not be concluded that the debugging phase was completed.

In this case, examination of the data logs for the three runs indicated a flaw in the experimental procedure and not in the instrumentation. In starting the experimental facility the morning of a run (which took from 4 to 8 h to complete), the laboratory barometer was read and that value of barometric pressure entered into the computer program that was used for control and data acquisition. That same value of barometric pressure was used in data reduction. On January 19, 1988, a weather front moved through the area, causing severe thunderstorms, tornado warnings, and large excursions in barometric pressure over a time period of a few hours. Numerical experiments with the data reduction program and the 01/19/88 data showed that the discrepancy in the St data could be explained by the change in barometric pressure that occurred between the time the barometer was read and the time the rest of the variable values were read when steady state was achieved several hours later.

This check at the first-order replication level made during the debugging phase identified a flaw in the experimental procedure that could cause unanticipated variations in the experimental data. This was remedied by a change in the experimental procedure requiring the values of *all* variables to be determined and recorded within a period of several minutes once a steady-state condition had been achieved.

Figure 5.20 shows the results of several separate Stanton number runs at $u_\infty = 12$ m/s. Once again it is apparent that these data sets do not support the estimate of a zero random uncertainty. The data clearly indicate a run-to-run scatter that cannot be termed negligible. Investigation showed that at low free-stream velocities ($u_\infty \leq 12$ m/s), the heat transfer coefficients are relatively low and the time constant of the test facility is thus increased. At these conditions the time constant of the test facility is large enough that the relatively long period variations in facility line voltage to the test plate heater circuits and in the temperature of the incoming makeup water for the heat exchanger loop affect the ability to hold a tight steady-state condition. These annoyances could be overcome with additional expenditures for power conditioning equipment and a water chiller system; however, the observed run-to-run scatter in Stanton number results was within acceptable limits, considering the research program objectives. (*Note*: The debugging phase of the program did not extend over a year-long period. The 08/26/88 data are from a check run made by a new graduate student.)

Examination of St data for several runs with $u_\infty = 12$ m/s produced a 95% confidence estimate of the random portion of uncertainty of about 3%. This is present because of system unsteadiness and not because of measurement uncertainties. The first-order replication-level check was again used to advantage in

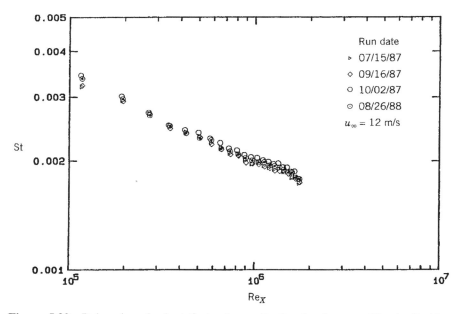

Figure 5.20 Debugging check at first-order replication level, $u_\infty = 12$ m/s. Problem indicated since $s_{St} \approx 0$ was expected.

identifying the effects of system unsteadiness and in estimating a new random
uncertainty contribution for low free-stream velocity runs.

Once a satisfactory conclusion had been reached at the first-order replication
level, checks at the Nth-order replication level were made. Since neither accepted
experimental data nor trustworthy theoretical results existed for the rough surface
conditions to be investigated in the experimental program, the only "truth" available with which comparisons could be made at the Nth-order replication level
was smooth surface St versus Re_x data that had been reported previously and
were widely accepted as valid. Such data had been reported by Reynolds, Kays,
and Kline [10] in 1958 for Reynolds numbers up to 3,500,000. These data are
shown in Figure 5.21 along with the power law correlation equation suggested
by Kays and Crawford [11]:

$$\text{St} = 0.0287(\text{Re}_x)^{-0.2}(\text{Pr})^{-0.4} \tag{5.76}$$

where Pr is the Prandtl number of the free-stream air. As can be seen, an interval of $\pm 5\%$ about Eq. (5.76) encloses roughly 95% of the data. If these data
are accepted as the standard, we are essentially assuming with 95% confidence
that the "true" Stanton numbers in the Reynolds number range from 100,000 to
3,500,000 lie in the $\pm 5\%$ band about Eq. (5.76)

Shown in Figure 5.22 is the qualification check at the Nth-order replication level. Data from the experiment are plotted along with uncertainty intervals
($\pm U_{\text{St}}$) for some representative data points. (The uncertainty interval on Re_x was

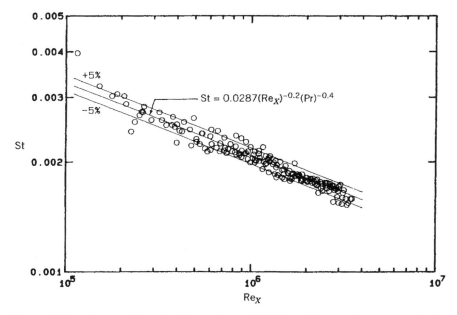

Figure 5.21   Classic data set of Reynolds et al. [10], widely accepted as a standard.

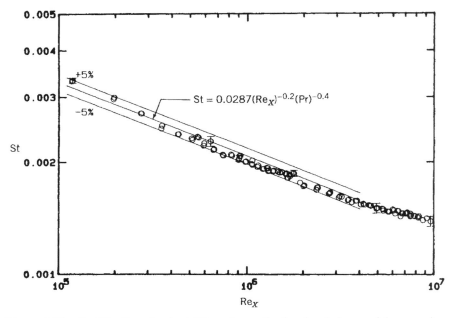

Figure 5.22 Qualification check at Nth-order replication level. Successful comparison of new data with the standard set.

less than $\pm 2\%$ and is therefore smaller than the width of the data symbols for the Re_x scale used in the figure.) In this case, the agreement is excellent in the sense that the new St data with the associated $\pm U_{St}$ intervals fall almost totally within the uncertainty band associated with the standard data. It was therefore concluded that a satisfactory qualification check for St at the Nth-order replication level had been made for $100{,}000 < Re_x < 3{,}500{,}000$. This then lent substantial credibility to the uncertainty estimates associated with the smooth-wall St data for $Re_x > 3{,}500{,}000$ shown in Figure 5.22 and also to the subsequent data taken with rough surfaces in the same apparatus and using the same instrumentation.

5-7 SOME ADDITIONAL CONSIDERATIONS IN EXPERIMENT EXECUTION

5-7.1 Choice of Test Points: Rectification

In the execution phase of an experimental program, the results are obtained for various "set" values of the measured variables. These set points should be chosen carefully so that the experiment will provide uniform information about the results over the full range of expected values. In this section we consider the concept of rectification and its use in test point spacing.

Once the test points are chosen, the order in which they are to be run must be decided. As we will see, a random order of test point settings rather than a sequential order is preferred.

The spacing of the test data points is achieved by choosing the value of the controlled variables that will be set. The values of the dependent variables are determined by measurement at each of the settings of the controlled variables. These test points should be chosen with careful thought before the experiment is begun. If the relationship between the dependent variable and the controlled variable is nonlinear, an equal spacing of the controlled variable settings over the range of interest will lead to an unequal spacing in the dependent variable. This may lead to more data points than needed in one range of the variables and too few points in another range. The idea of rectification is key in the choice of coordinate systems that should be considered when determining test point spacing.

Assume that we are dealing with an experiment that produces data of high precision—that is, groups of data pairs (x, y) that give a smooth curve with relatively little scatter when plotted as y versus x. With high-precision data, we try to use the coordinates that will rectify the data or cause the points to fall on a straight line. Note that here x and y might very well be nondimensional parameters consisting of several different measured variables.

When plotted in a straight line, the data can be represented by the equation

$$Y = aX + b \tag{5.77}$$

1. For data pairs (x, y) whose functional relationship is

$$y = ax + b \tag{5.78}$$

we see that $Y = y$ and $X = x$ and we obtain a straight line by plotting y versus x on linear–linear coordinates.

2. For data pairs (x, y) whose functional relationship is

$$y = cx^d \tag{5.79}$$

we can take the log of (5.79) and obtain

$$\log y = \log c + d \log x \tag{5.80}$$

so that, comparing with Eq. (5.77), we have $Y = \log y$, $X = \log x$, $b = \log c$, and $a = d$. Therefore, if we plot $\log y$ versus $\log x$ on linear–linear coordinates, we will obtain a straight line. The same result can be obtained by plotting y versus x on log–log coordinates.

3. For data pairs (x, y) whose functional relationship is

$$y = ce^{dx} \tag{5.81}$$

we can take the natural logarithm (ln) of Eq. (5.81) and obtain

$$\ln y = \ln c + dx \qquad (5.82)$$

so that, comparing with Eq. (5.77), we have $Y = \ln y$, $X = x$, $a = d$, and $b = \ln c$. Therefore, if we plot $\ln y$ versus x on linear–linear coordinates, we will obtain a straight line. Plotting y versus x on semilog coordinates also produces a straight line.

These three cases are the most common that occur in engineering. How do we use this information? If we know the general functional relationship between the variables before running an experiment, we should plan to present the data in the coordinates that rectify the data.

In many cases, the general form of the functional relationship between the variables is known, and this can be used to determine test point spacing. The following two examples illustrate this point.

Example 5.2   In a certain experiment X is the controlled variable and Y is the result. Previous results in similar physical cases indicate a probable functional relationship of the form

$$Y = aX^b \qquad (5.83)$$

We intend to curve fit the data to obtain the "best" values of the constants a and b. What spacing of X values should be used? Note that we must choose the data point spacing before we know the values of a and b.

Solution   If we take the logarithm of both sides of (5.83), we obtain

$$\log Y = \log a + b \log X \qquad (5.84)$$

From Eq. (5.84) we find that if we consider $\log Y$ and $\log X$ as our variables, we have a linear relationship irrespective of the exponent of X. We can therefore use equal increments of $\log X$ as our set points to obtain equally spaced values of $\log Y$.

Example 5.3   Same statement as above, except the probable relationship is of the form

$$Y = ae^{-bX} \qquad (5.85)$$

Solution   Again taking logarithms, we obtain

$$\ln Y = \ln a - bX \qquad (5.86)$$

and we see that using equally spaced values of X and considering $\ln Y$ as the dependent variable will lead to evenly spaced data points.

Another factor that should be taken into account when determining test point spacing is the expected uncertainty in the result over the range of interest. If uncertainty analysis has indicated a higher random uncertainty contribution to the uncertainty in one portion of the range than another, one might want to take more points in the region in which the uncertainty is highest, remembering that

$$s_{\bar{y}} = \frac{s_y}{\sqrt{N}} \tag{5.87}$$

Therefore, as a rule of thumb, four times the number of points gives a twofold decrease in the random uncertainty contribution to the uncertainty.

5-7.1.1 Example of Use of Rectification An experimental water flow facility to determine pipe friction factors for various Reynolds numbers is shown in Figure 5.23. The pump speed is set so that the desired Reynolds number

$$\mathrm{Re} = \frac{4\rho Q}{\pi \mu D} \tag{5.88}$$

is achieved, and then the friction factor

$$f = \frac{\pi^2 D^5 \Delta P}{32 \rho Q^2 \Delta X} \tag{5.89}$$

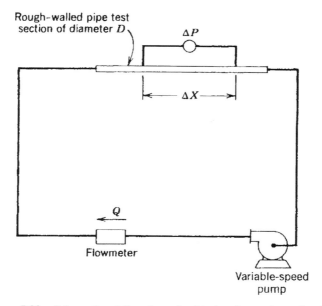

Figure 5.23   Schematic of flow loop for friction factor determination.

is determined from measurements of ΔP and Q. It is known from previous experiments that for a given pipe roughness the friction factor is a function of the Reynolds number.

Such data are traditionally presented in a $\log f - \log$ Re format—recall the famous Moody diagram from basic fluid mechanics. Shown in Figure 5.24 are the results of testing two test sections with rough walls for 14 Reynolds numbers from 13,000 to 611,000, where the wall roughness in test a was greater than that in test b. The upper data set was run with set points equispaced in Re, and the lower set was run with set points equispaced in log Re. The advantage of choosing set points with equal spacing in the coordinates of data presentation is obvious.

The determination of the Re set points for equal spacing in log Re is as follows:

$$\left. \begin{array}{l} \text{Re} = 611{,}000 \rightarrow \log \text{Re} = 5.786 \\ \text{Re} = 13{,}000 \rightarrow \ \log \text{Re} = 4.114 \end{array} \right\} \ \Delta = 1.672$$

Divide Δ by $N - 1 = 13 \rightarrow 0.1286$. Therefore,

$$(\log \text{Re})_i = (\log \text{Re})_{i-1} + 0.1286 \quad (\text{for } i = 2, \ldots, 14)$$

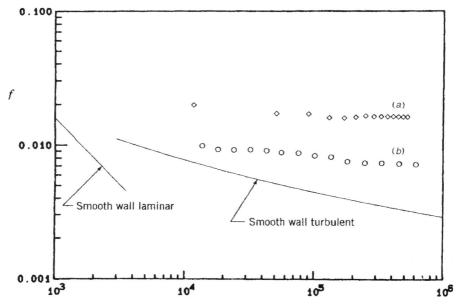

Figure 5.24 Friction factor versus Reynolds number data for two rough wall pipes: (a) points equispaced in Re; (b) points equispaced in log Re.

i	log Re	Re
1	4.1140	13,002
2	4.2426	17,482
3	4.3712	23,507
4	4.4998	31,608
5	4.6284	42,501
6	4.7570	57,147
7	4.8856	76,842
8	5.0142	103,323
9	5.1428	138,931
10	5.2714	186,810
11	5.4000	251,189
12	5.5286	337,754
13	5.6572	454,151
14	5.7858	610,661

In this example we have illustrated the use of rectification to determine test point spacing for the case in which both the dependent variable (f) and the independent variable (Re) are results determined from measured quantities (Q, D, ΔP, and ΔX) and property values (μ and ρ). These two results are appropriate dimensionless groups for presenting this type of data. By arranging the measured values and the property data into dimensionless groups, we have reduced the number of variables that must be varied in the experiment.

To set each of the Reynolds numbers specified as test points, we can vary the pipe diameter (D), the flow rate (Q), or the fluid itself (ρ and μ). In an actual experiment we would probably keep the fluid and the pipe diameter fixed and vary the flow rate. These results would then be general and would apply for any other flow situations that had the same Reynolds number and nondimensional roughness.

The appropriate dimensionless groups should be considered in any test plan. Many references are available on dimensional analysis. In general, its use will make the experiment easier to conduct and will make the results more universally applicable.

5-7.2 Choice of Test Sequence

There are many arguments for choosing a random order of test point settings rather than a sequential order (from lowest to highest value, or vice versa). Among these are [12]

1. *Natural effects*. Uncontrolled (or "extraneous") variables might have a small but nevertheless measurable effect on the results and may follow a

general trend during the test. Examples are relative humidity and barometric pressure.

2. *Human activities*. The operator/data recorder may become more proficient during the test or may become bored and careless.

3. *Hysteresis effects*. Instruments may read high if the previous reading was higher and low if the previous reading was lower.

When possible, random sequences should be used when conducting an experiment. As an example, suppose that a person is measuring the velocity V of a point on the edge of spinning disk of radius $r = 10$ cm as a function of ω. The person is unaware of the relationship

$$V = r\omega \qquad (5.90)$$

Suppose further that the only errors are systematic errors due to hysteresis in the velocity measurement system, which reads 5% low if the set point is approached from below and 5% high if the set point is approached from above.

Shown in Figure 5.25 are one data set taken in a sequentially increasing manner and one data set taken in a sequentially decreasing manner compared with the true relationship. It is obvious that the "best" line through either data set does not give a good picture of the true relationship.

Shown in Figure 5.26 is a data set with the points taken in a random order, with some points approached from above and some from below. Although the

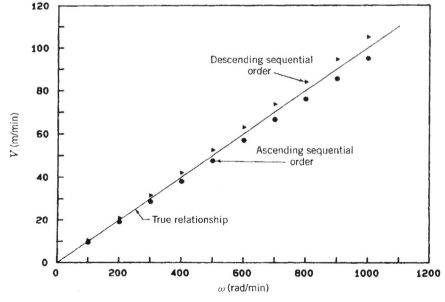

Figure 5.25 Data taken in sequential order and including 5% hysteresis effect.

points in this data set scatter more than the sets in Figure 5.25, they scatter about the true relationship and thus give a better idea of the true system behavior.

In a general experiment in which the result r is a function of several variables,

$$r = f(x, y, z) \tag{5.91}$$

the test plan will consist of holding all of the independent variables but one fixed and varying that one over its full range. The process is then repeated for the other independent variables. This approach is referred to as a classical test plan and is the common method used for most engineering experiments [12].

5-7.3 Relationship to Statistical Design of Experiments

Statistical design of experiments (DOE) is a process of experimental planning such that the data obtained are suitable for a statistical analysis. Statistical tests [such as analysis of variance (ANOVA)] are then applied to the data to determine whether or not possible functional dependencies of the experimental result on process variables are of statistical significance or not. Typically, a regression (using the dependencies found to be significant) is then performed to obtain a mathematical relationship that represents the experimental data.

This type of experimental approach was originally developed and applied in the field of quality control and process optimization. Many quality control

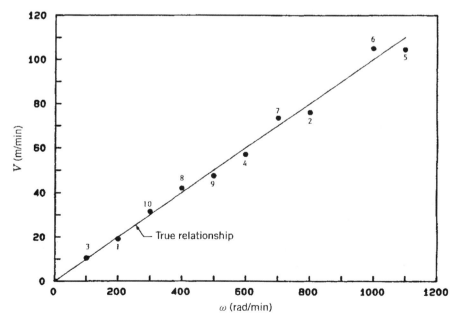

Figure 5.26   Data taken in random order and including 5% hysteresis effect.

and product development experiments are based on the assumption that the system/process is sensitive to certain factors but is less sensitive to the interaction of these factors [13]. This assumption forms the foundation of the Taguchi approach which was much in vogue in the 1990s [14].

Such approaches have found limited usefulness in testing of engineering systems and processes governed by systems of partial differential equations and in which a result $r(x, y)$ may depend on the independent variables or other results in ways such as $x^a y^b$—consider heat transfer data correlations such as $Nu = a(Re)^b (Pr)^c$, for instance.

Deaconu and Coleman [15] reported results of a study in which Monte Carlo simulations were used to examine the effectiveness of two DOE ANOVA methods combined with F-test statistics in identifying the correct terms in postulated regression models for a variety of experimental conditions. It was concluded that the ability of the statistical approaches to identify the correct models varied so drastically, depending on experimental conditions, that it seems unlikely that arbitrarily choosing a method and applying it will lead to identification of the effects that are significant with a reasonable degree of confidence. It was concluded that before designing and conducting an experiment based on DOE methods one should use simulations of the proposed experiment with postulated truths in order to determine which DOE approach, if any, will identify the correct model from the experimental data with an acceptable level of confidence.

5-7.4 Use of Balance Checks

Balance checks are an application of the basic physical conservation laws (energy, mass, electrical current, etc.) to an experiment. The reasons for using balance checks are twofold. First, in the debugging/qualification phase the application of a conservation law can help determine if errors exist that have not been taken into account. If a balance check is not satisfied within the estimated uncertainty interval (Nth order) for the measurements, the debugging/qualification phase of the experiment cannot be assumed to be complete. Second, the use of balance checks that have been shown to be satisfied in the qualification phase can be useful in monitoring the execution phase of an experiment. If, for instance, a balance check suddenly fails to be satisfied, this can indicate an unrecognized change in instrument calibration or the process itself.

Balance checks may require more measurements than are needed by the basic experiment itself. Additional measurements must then be planned for in the design phase of the experiment.

5-7.4.1 Application to a Flow System   Consider the flow system in Figure 5.27. Three flowmeters are installed to measure the three mass flow rates m_1, m_2, and m_3. If all the measurements were perfect, conservation of mass would say that

$$m_3 = m_1 + m_2 \tag{5.92}$$

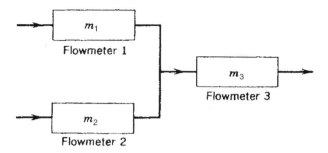

Figure 5.27 Flow system schematic for balance check example.

However, uncertainties in the measurements (and possible leaks) make this *extremely* unlikely. How closely must Eq. (5.92) be satisfied before we can conclude that the balance check is satisfactory?

Take the quantity z as the mass balance parameter

$$z = m_3 - m_1 - m_2 \tag{5.93}$$

Now the question becomes: How close to zero must z be before we can say that the mass balance is satisfied? This must be answered if we are to use this mass balance to check our experiment in the debugging phase or monitor the experiment during the execution phase.

Considering z as an experimental result, the TSM propagation expressions are

$$s_z^2 = \left(\frac{\partial z}{\partial m_1}\right)^2 s_{m_1}^2 + \left(\frac{\partial z}{\partial m_2}\right)^2 s_{m_2}^2 + \left(\frac{\partial z}{\partial m_3}\right)^2 s_{m_3}^2 \tag{5.94}$$

and

$$
\begin{aligned}
b_z^2 = {} & \left(\frac{\partial z}{\partial m_1}\right)^2 b_{m_1}^2 + \left(\frac{\partial z}{\partial m_2}\right)^2 b_{m_2}^2 + \left(\frac{\partial z}{\partial m_3}\right)^2 b_{m_3}^2 \\
& + 2\left(\frac{\partial z}{\partial m_1}\right)\left(\frac{\partial z}{\partial m_2}\right) b_{m_1 m_2} \\
& + 2\left(\frac{\partial z}{\partial m_1}\right)\left(\frac{\partial z}{\partial m_3}\right) b_{m_1 m_3} \\
& + 2\left(\frac{\partial z}{\partial m_2}\right)\left(\frac{\partial z}{\partial m_3}\right) b_{m_2 m_3}
\end{aligned} \tag{5.95}
$$

where, as usual, the $b_{m_i m_j}$ are the portions of the systematic uncertainties in m_i and m_j that are from the same source and therefore account for the correlated errors. As

$$\frac{\partial z}{\partial m_1} = \frac{\partial z}{\partial m_2} = -1 \tag{5.96}$$

and

$$\frac{\partial z}{\partial m_3} = +1 \tag{5.97}$$

Eqs. (5.94) and (5.95) become

$$s_z^2 = s_{m_1}^2 + s_{m_2}^2 + s_{m_3}^2 \tag{5.98}$$

and

$$b_z^2 = b_{m_1}^2 + b_{m_2}^2 + b_{m_3}^2 + 2b_{m_1 m_2}$$
$$- 2b_{m_1 m_3} - 2b_{m_2 m_3} \tag{5.99}$$

and the overall 95% confidence uncertainty in z is given by

$$U_z^2 = 2^2(b_z^2 + s_z^2) \tag{5.100}$$

For the balance check to be satisfied at a 95% level of confidence, the following must be true:

$$|z| \leq U_z \tag{5.101}$$

To investigate the behavior of U_z for various circumstances, assume for purposes of this example that all s values and b values are equal to 1 kg/h and the $b_{m_i m_j}$ are equal to $(b_{m_i})(b_{m_j})$ in the instances in which they are nonnegligible:

1. Random uncertainty dominated—all systematic uncertainties negligible:

$$U_z = 2s_z = 2(s_{m_1}^2 + s_{m_2}^2 + s_{m_3}^2)^{1/2}$$
$$= 2(1 + 1 + 1)^{1/2} = 3.5 \text{ kg/h}$$

Therefore, $|z| \leq 3.5$ kg/h for the mass balance to be satisfied.

2. Random uncertainties negligible—all systematic errors uncorrelated:

$$U_z = 2b_z = 2(b_{m_1}^2 + b_{m_2}^2 + b_{m_3}^2)^{1/2}$$
$$= 2[(1) + (1) + (1)]^{1/2} = 3.5 \text{ kg/h}$$

Therefore, $|z| \leq 3.5$ kg/h for the mass balance to be satisfied.

3. Random uncertainties negligible—systematic errors in m_1 and m_2 correlated:

$$U_z = 2b_z = 2(b_{m_1}^2 + b_{m_2}^2 + b_{m_3}^2 + 2b_{m_1 m_2})^{1/2}$$
$$= 2[(1) + (1) + (1) + (2)(1)(1)]^{1/2} = 4.5 \text{ kg/h}$$

Therefore, $|z| \leq 4.5$ kg/h for the mass balance to be satisfied. This case could occur if all systematic error in m_1 and m_2 was from calibration and both were calibrated against the same standard, while meter 3 was calibrated against a different standard.

4. Random uncertainties negligible—systematic errors in m_1, m_2, and m_3 correlated:

$$U_z = 2b_z$$
$$= 2(b_{m_1}^2 + b_{m_2}^2 + b_{m_3}^2 + 2b_{m_1 m_2} - 2b_{m_1 m_3} - 2b_{m_2 m_3})^{1/2}$$
$$= 2[(1) + (1) + (1) + (2)(1)(1) - (2)(1)(1) - (2)(1)(1)]^{1/2}$$
$$= 2.0 \text{ kg/h}$$

Therefore, $|z| \leq 2.0$ kg/h for the mass balance to be satisfied. This case could occur if all systematic errors were from calibration and all three meters were calibrated against the same standard.

5. Random and systematic uncertainties important—systematic errors in m_1 and m_2 correlated:

$$U_z = 2(s_z^2 + b_z^2)^{1/2}$$
$$= 2(s_{m_1}^2 + s_{m_2}^2 + s_{m_3}^2 + b_{m_1}^2 + b_{m_2}^2 + b_{m_3}^2 + 2b_{m_1 m_2})^{1/2}$$
$$= 2[(1) + (1) + (1) + (1) + (1) + (1) + (2)(1)(1)]^{1/2}$$
$$= 5.7 \text{ kg/h}$$

Therefore, $|z| \leq 5.7$ kg/h for the mass balance to be satisfied.

In summary, for this example, we have seen the general approach to formulating a balance check using uncertainty analysis principles. We also note the possibility of correlated systematic errors playing a major role in determining the degree of closure required for the balance to be satisfied.

5-7.5 Use of a Jitter Program

A jitter program can be an extremely useful tool and should be implemented in the debugging/qualification and execution phases of almost every experimental program. In a jitter program, the data reduction computer program is treated as a subroutine that is successively iterated so that all of the partial derivatives necessary in uncertainty analysis are calculated using finite-difference approximations (as discussed in Section 3-1.5). Moffat [16–18] introduced this idea for propagating general uncertainty intervals into a result. Here, we extend the idea to include the separate propagation of systematic and random uncertainties into a result and also to include the effects of correlated systematic errors.

The biggest advantage in using a jitter program is that whenever a change in the data reduction program is made, the uncertainty analysis program is automatically updated since the data reduction program itself is used to generate the partial derivative values. In complex experiments with extended debugging/qualification phases, this is of great importance. The ease with which the jitter program can be implemented, however, practically dictates its use even in experiments with very simple data reduction programs.

The jitter program procedure is outlined in Figure 5.28 for a case in which the experimental result r is a function of K variables X_i. The X_i values include not only the variables that are measured but also those (such as emissivity, for example) whose values are usually found from a reference source. These latter variables should be included in the data reduction program as named variables, not as a single numerical value, as they must be perturbed in the jitter program so that a partial derivative can be calculated.

As seen in the figure the inputs required are the data itself (the values of the X_i), the estimates of s and b for each variable, the estimates of the correlated systematic error terms (if any), and specification of the amount (ΔX_i) each variable is to be perturbed in the finite-difference approximation of the partial derivatives. This is used $K + 1$ times to compute the results r_0 and a perturbed result $r_{X_i + \Delta X_i}$ for each variable, and the finite-difference approximations to the partial derivatives are computed together with the terms for the s_r and b_r propagation equations. At this point, s_r, b_r, and u_r are calculated and each term contributing to s_r^2 and b_r^2 is divided by u_r^2 to give the uncertainty percentage contribution.

A suggested output format is shown in the figure. The result r_0 is (obviously) output and u_r, b_r, and s_r are output both in dimensional form and as a percentage of r_0. The percentage output form is especially useful for allowing a quick comprehension of the importance of the uncertainty components when the result and/or u_r, b_r, and s_r are either very large or very small numbers relative to unity.

The second part of the output has been found to be very useful by the authors. The array of normalized values of the individual contributors to b_r^2 and s_r^2 contains numbers that add to 100, with the correlated terms having the possibility of being negative as well as positive. One can quickly glance over this array and discern which contributors are important and which are negligible. This is particularly important to track in experiments that cover a broad range of conditions. In such cases a contributor might well be negligible in one regime but be dominant in another.

5-7.6 Comments on Transient Testing

We briefly discussed types of transient tests in Section 1-3.3. In transient tests, the process of interest varies with time and there are some unique circumstances involved in obtaining multiple measurements of a variable at a given condition. As an example of one type of transient test, consider that the thrust of a rocket engine at $t = 0.1$ s during the startup transient is the quantity of interest. To obtain a sample population of measurements of thrust at $t = 0.1$ s, the engine must be tested again, and again, and again. This type of test has obvious similarities to the sample-to-sample category, and even more so if multiple engines of the same design are considered as part of the sample population of measurements.

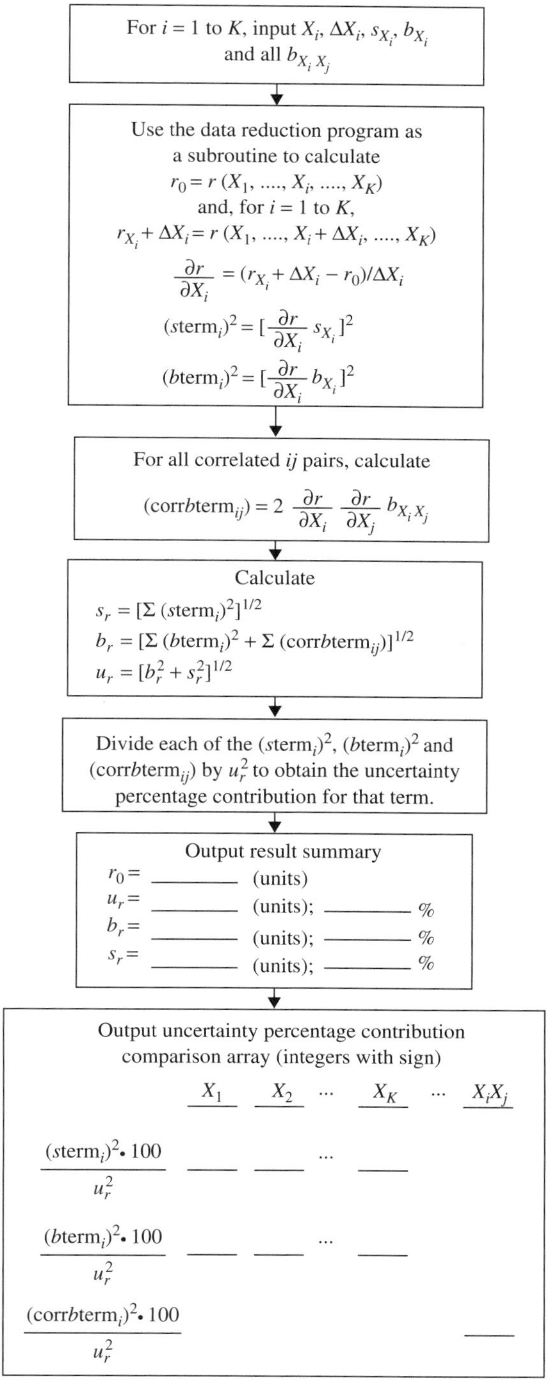

For $i = 1$ to K, input X_i, ΔX_i, s_{X_i}, b_{X_i} and all $b_{X_i X_j}$

Use the data reduction program as a subroutine to calculate
$r_0 = r(X_1, ..., X_i, ..., X_K)$
and, for $i = 1$ to K,
$r_{X_i} + \Delta X_i = r(X_1, ..., X_i + \Delta X_i, ..., X_K)$

$$\frac{\partial r}{\partial X_i} = (r_{X_i} + \Delta X_i - r_0)/\Delta X_i$$

$$(sterm_i)^2 = [\frac{\partial r}{\partial X_i} s_{X_i}]^2$$

$$(bterm_i)^2 = [\frac{\partial r}{\partial X_i} b_{X_i}]^2$$

For all correlated ij pairs, calculate

$$(corrbterm_{ij}) = 2 \frac{\partial r}{\partial X_i} \frac{\partial r}{\partial X_j} b_{X_i X_j}$$

Calculate
$$s_r = [\Sigma (sterm_i)^2]^{1/2}$$
$$b_r = [\Sigma (bterm_i)^2 + \Sigma (corrbterm_{ij})]^{1/2}$$
$$u_r = [b_r^2 + s_r^2]^{1/2}$$

Divide each of the $(sterm_i)^2$, $(bterm_i)^2$ and $(corrbterm_{ij})$ by u_r^2 to obtain the uncertainty percentage contribution for that term.

Output result summary
$r_0 =$ _____ (units)
$u_r =$ _____ (units); _____ %
$b_r =$ _____ (units); _____ %
$s_r =$ _____ (units); _____ %

Output uncertainty percentage contribution comparison array (integers with sign)

	X_1	X_2	$\cdots$	X_K	$\cdots$	$X_i X_j$
$\dfrac{(sterm_i)^2 \cdot 100}{u_r^2}$	___	___	...	___		
$\dfrac{(bterm_i)^2 \cdot 100}{u_r^2}$	___	___	...	___		
$\dfrac{(corrbterm_i)^2 \cdot 100}{u_r^2}$						___

Figure 5.28 Schematic of jitter program procedure.

A second type of transient test is that when the process is periodic. Consider measuring the temperature at a point within a cylinder of an internal combustion engine operating at a steady condition. This type of test has some obvious similarities to the timewise steady category. Multiple measurements of temperature at the "same" operating point can be obtained by making a measurement at $30°$ before top dead center, for example, over a number of repetitions of the periodic process.

Measurements in a transient test typically have elemental error sources in addition to those normally present in either a timewise steady test or a sample-to-sample test. Imperfect dynamic response of the measurement system can result in errors in the magnitude and phase of the measured quantity. An overview of the dynamic response of instrument systems and these types of errors is presented in Appendix F.

REFERENCES

1. Hudson, S. T., Bordelon, W. J., and Coleman, H. W., "Effect of Correlated Precision Errors on the Uncertainty of a Subsonic Venturi Calibration," *AIAA Journal*, Vol. 34, No. 9, Sept. 1996, pp. 1862–1867.

2. Coleman, H. W., and Lineberry, D. L., "Proper Estimation of Random Uncertainties in Steady State Testing," *AIAA Journal*, Vol. 44, No. 3, 2006, pp. 629–633.

3. Moffat, R. J., "Describing the Uncertainties in Experimental Results," *Experimental Thermal and Fluid Science*, Vol. 1, Jan. 1998, pp. 3–17.

4. U.S. National Bureau of Standards (NBS), *Tables of Thermal Properties of Gases*, NBS Circular 564, NBS, Washington, DC, 1955.

5. Brown, K. K., Coleman, H. W., Steele, W. G., and Taylor, R. P., "Evaluation of Correlated Bias Approximations in Experimental Uncertainty Analysis," *AIAA Journal*, Vol. 34, No. 5, May 1996, pp. 1013–1018.

6. American Society of Heating, Refrigerating, and Air-Conditioning Engineers (ASHRAE), *Fundamentals, ASHRAE Handbook*, ASHRAE, Atlanta, GA, 1997.

7. Coleman, H. W., Steele, W. G., and Taylor, R. P., "Implications of Correlated Bias Uncertainties in Single and Comparative Tests," *Journal of Fluids Engineering*, Vol. 117, Dec. 1995, pp. 552–556.

8. Chakroun, W., Taylor, R. P., Steele, W. G., and Coleman, H. W., "Bias Error Reduction Using Ratios to Baseline Experiments—Heat Transfer Case Study," *Journal of Thermophysics and Heat Transfer*, Vol. 7, No. 4, Oct.– Dec. 1993, pp. 754–757.

9. Coleman, H. W., Hosni, M. H., Taylor, R. P., and Brown, G. B., "Using Uncertainty Analysis in the Debugging and Qualification of a Turbulent Heat Transfer Test Facility," *Experimental Thermal and Fluid Science*, Vol. 4, 1991, pp. 673–683.

10. Reynolds, W. C., Kays, W. M., and Kline, S. J., *Heat Transfer in the Turbulent Incompressible Boundary Layer*, Parts I, II, and III, NASA memos 12-1-58W, 12-2-58W, and 12-3-58W, 1958.

11. Kays, W. M., and Crawford, M. E., *Convective Heat and Mass Transfer*, McGraw-Hill, New York, 1980.

12. Schenck, H., *Theories of Engineering Experimentation*, 3rd ed., McGraw-Hill, New York, 1979.

13. Hicks, C. R., *Fundamental Concepts in the Design of Experiments*, 3rd ed., Holt, Rinehart & Winston, New York, 1982.

14. Montgomery, D. C., *Design and Analysis of Experiments*, 4th ed., Wiley, New York, 1997.

15. Deaconu, S., and Coleman, H. W., "Limitations of Statistical Design of Experiments Approaches in Engineering Testing," *Journal of Fluids Engineering*, Vol. 122, No. 2, 2000, pp. 254–259.

16. Moffat, R. J., "Contributions to the Theory of Single-Sample Uncertainty Analysis," *Journal of Fluids Engineering*, Vol. 104, June 1982, pp. 250–260.

17. Moffat, R. J., "Using Uncertainty Analysis in the Planning of an Experiment," *Journal of Fluids Engineering*, Vol. 107, June 1985, pp. 173–178.

18. Moffat, R. J., "Describing the Uncertainties in Experimental Results," *Experimental Thermal and Fluid Science*, Vol. 1, Jan. 1988, pp. 3–17.

PROBLEMS

5.1 The elemental systematic standard uncertainties affecting the measurement of temperature with a particular thermistor probe are estimated as $0.2°C$ from the calibration standard, $0.35°C$ due to temperature nonuniformity of the constant-temperature bath used during calibration, $0.5°C$ from use of a curve-fit equation to represent the calibration data, and $0.2°C$ from installation effects. What is an appropriate estimate for the systematic standard uncertainty of a temperature measurement made using this probe?

5.2 Compare the digitizing 95% confidence systematic uncertainty for a signal of $±0.5$ V or less for 8-, 12-, and 16-bit A/D converter systems for calibrated ranges of $±5$ V and $±0.5$ V.

5.3 Consider the calibration of an instrument system consisting of a transducer connected to a signal conditioner with a built-in digital meter. The meter is calibrated for a range of $±100$ mV with a least digit of 1 mV and the meter uses an 8-bit A/D converter. If the input signal to the meter is constant at 2 mV and if the calibration standard error is negligible, estimate the random and systematic uncertainties for this reading. (Be sure to consider the digitizing systematic uncertainty.)

5.4 During a test, the reading on a digital meter varies randomly from 1.22 to 1.26 V. The meter has a calibrated range of 0 to 10 V and uses a 12-bit A/D converter. The manufacturer specifies that the 95% confidence accuracy of the meter is $±(2\%$ of the reading $+1$ times the least scale division). Estimate the systematic and random uncertainties of the reading.

5.5 Consider the experimental data for the specific heat of air as presented in Figure 5.12. What would be a reasonable estimate for the fossilized 95% confidence systematic uncertainty associated with use of tabulated values of c_p in the range 250 to 300 K? In the range 450 to 550 K? What assumptions do you make in arriving at these estimates?

5.6 Consider the experimental data for the thermal conductivity of air as presented in Figure 5.13. What would be a reasonable estimate for the fossilized 95% confidence systematic uncertainty associated with use of tabulated values of k in the temperature range close to 300 K? In the range close to 500 K? What assumptions do you make in arriving at these estimates?

5.7 In a testing program to investigate the effectiveness of different lumber-drying procedures, a load of lumber is weighed on scale Y, taken to a nearby kiln for drying, and then taken to scale Z, where it is weighed. The amount of moisture lost during the drying operation is determined by subtracting the second weight from the first. The only systematic uncertainties of significance are from the calibrations of the two scales. Determine a reasonable estimate for the systematic standard uncertainty of the moisture determination if

$$b_{cal,Y} = 150 \text{ lbf} \qquad b_{cal,Z} = 175 \text{ lbf}$$

5.8 For the situation in problem 5.7 what would be the effect on the systematic uncertainty of the moisture determination if the two scales were calibrated using the same truck as a reference standard but the weight of the truck is only known with a systematic standard uncertainty of 500 lbf?

5.9 The rate at which energy is removed from an air-cooled system is being determined by measuring the air inlet and outlet temperatures (T_i and T_0) and the air mass flow rate and using

$$\dot{E} = \dot{m}c_p(T_0 - T_i)$$

where c_p is the specific heat of air at constant pressure and is evaluated at an average temperature. For the operating condition of interest, nominal inlet and outlet temperatures of about 300 and 320 K are expected. A fossilized systematic standard uncertainty for c_p has been estimated as 0.5%, and the manufacturer of the mass flowmeter guarantees "an absolute accuracy of 0.5% of reading." The contributions to the systematic standard uncertainties for the temperature measurements include a 0.2 K and 0.4 K nonuniformity effect at the inlet and outlet, respectively, and calibration systematic standard uncertainties of 0.5 K for both temperature probes (which share no errors). Estimate the systematic standard uncertainty for the result, $\dot{E}$

5.10 For the situation in problem 5.9 could there be any improvement by calibrating the two temperature probes against a thermometer with a systematic standard uncertainty of 1.0 K? What would the new systematic standard uncertainty in $\dot{E}$ be?

5.11 For the situation in problem 5.9 replication of the test over a long time period indicates an inherent unsteadiness that results in 95% confidence random uncertainties of 0.8% for $\dot{m}$ and 0.8 K for each of the temperatures.

What is the random standard uncertainty for the result, $\dot{E}$? What is the 95% confidence estimate for the overall uncertainty in $\dot{E}$? How does taking advantage of the correlated systematic error effect (problem 5.10) change the overall uncertainty in $\dot{E}$?

5.12 A heat exchanger manufacturer is testing two designs ("f" and "g") of an air-to-water heat exchanger. The two heat exchanger cores have different fin configurations on the air side. A typical core design is shown schematically below.

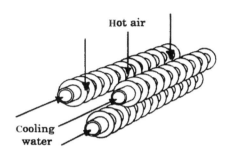

To evaluate the designs, the rate of heat transfer to the water is determined for a given set point using

$$q = mc\left(\frac{T_2 + T_3 + T_4}{3} - T_1\right)$$

where q is the rate of heat transfer from the hot air to the cooling water; m is the mass flow rate of the water; c is the constant-pressure specific heat; T_2, T_3, and T_4 are measured by three temperature probes at different positions in a cross section in the water outlet header; and T_1 is the water temperature in the inlet header as measured by a single probe.

The data from the two tests are as follows:

	Configuration f	Configuration g	Systematic Standard Uncertainty, b
m (kg/s)	0.629	0.630	0.25% of reading
c(kJ/kg · K)	4.19	4.19	0.30%
$T_1(^\circ\text{C})$	15.7	15.4	See below
$T_{2,3,4}(^\circ\text{C})$	52.8, 53.1, 52.6	51.4, 51.6, 51.3	See below

The significant elemental systematic uncertainties for the temperature probes are $b_{\text{std}} = 0.3^\circ\text{C}$ and the b_{install} equal 0.1°C for the inlet and 0.2°C for the outlet. Using the same test facility and instrumentation, previous tests at about the same set point on a core design similar to those of f and g indicated a random standard uncertainty of roughly 0.4% of the value of q determined.

The better design is determined by comparing q_f and q_g for "identical" test conditions and determining which q is larger. Find the uncertainty which should be associated with the determination of

$$\Delta q = q_f - q_g$$

Perform an uncertainty analysis using the MCM and compare with an uncertainty analysis using the TSM. Consider a situation in which the four temperature probes are not calibrated against the same standard and also a situation in which they are calibrated against the same standard.

What is your recommendation relative to the merits of design f versus design g?

5.13 For a centrifugal pump, the power input required, p, for a given pump speed N is governed by

$$p = CN^3$$

A range of speeds from 500 to 2000 rpm is being investigated, and six points are to be run (including the endpoints). A linear representation with equal spacing of points is desired. How should the data be presented and what N values should be run?

5.14 The attenuation of a radiation beam as it passes through a medium can be expressed as

$$I = I_0 e^{-\beta x}$$

where I_0 is the intensity of the beam entering the medium and I is the intensity at a depth x. Assume β is a parameter that depends on the type of radiation and the medium. For a specific medium in which β will probably be a constant, it is planned to measure the intensity of gamma rays at different depths where x will vary from 0.1 to 2 m. Seven points are to be run (including the endpoints) and a linear representation with equally spaced points is desired. How should the data be presented, and what x locations should be used?

5.15 The heat transfer coefficient for freely falling liquid droplets can be expressed as

$$\text{Nu} = a + b\text{Re}^c$$

where a, b, and c are constants and Nu and Re are the Nusselt and Reynolds numbers, respectively. If the Reynolds number ranges up to 350, what Re values should be considered for test measurements if 10 data points are to be obtained? How should these data be represented graphically?

5.16 A heat exchanger, which uses water to cool the oil used in a transmission, is being tested to determine its effectiveness. An energy balance on the heat exchanger (assuming negligible energy loss to the environment) yields

$$\dot{m}_0 c_0 (T_{0,i} - T_{0,o}) = \dot{m}_w c_w (T_{w,o} - T_{w,i})$$

where
$\dot{m}_0 =$ mass flow rate of oil
$c_0 =$ specific heat of oil
$T_{0,i} =$ oil temperature at oil inlet
$T_{0,o} =$ oil temperature at oil outlet
$\dot{m}_w =$ mass flow rate of water
$c_w =$ specific heat of water
$T_{w,o} =$ water temperature at water outlet
$T_{w,i} =$ water temperature at water inlet

Assume that the random errors are negligible relative to systematic errors. The mass flow rate measurements (which share no errors) have 95% confidence systematic uncertainties of 1% of reading, and the values of specific heat have fossilized 95% confidence systematic uncertainties estimated as 1.5%. All temperature measurements have an elemental 95% confidence systematic uncertainty contribution of 1.0°C due to the nonuniformity of fluid temperatures across the flow inlets and outlets. All temperature measurements also have calibration 95% confidence systematic uncertainties of 1.5°C.

For nominal conditions such that

$$T_{0,i} - T_{0,o} = 40°C \qquad T_{w,o} - T_{w,i} = 30°C$$

determine the bounds within which an energy balance should be expected to close for (a) all systematic errors uncorrelated and (b) all calibration systematic errors in the temperature sensors correlated.

5.17 In a particular experiment, the dc electrical power supplied to a circuit component must be measured. A dc watt transducer that the manufacturer says can measure the power to within 0.4% of reading is to be used. As a check during debugging, the voltage drop (E), current (I), and resistance (R) of the component are to be measured with a multimeter that has uncertainty specifications of 0.1% for E, 0.3% for R, and 0.75% for I. Nominal values are 1.7 A, 12 Ω, and 20.4 V. Denote W_1 as the power indicated by the meter output, $W_2 = EI$, and $W_3 = I^2 R$. Assume all random uncertainties negligible. To consider the balance check satisfied, how closely should W_1 and W_2 agree? W_1 and W_3? W_2 and W_3? W_1, W_2, and W_3? (all uncertainties are 95% confidence).

6

VALIDATION OF SIMULATIONS

6-1 INTRODUCTION TO VALIDATION METHODOLOGY

Over the past several decades, advances in computing power, modeling approaches, and numerical solution algorithms have increased the ability of the scientific and engineering community to simulate real-world processes to the point that it is realistic for predictions from surprisingly detailed simulations to be used to replace much of the experimentation that was previously necessary to develop designs for new systems and bring them to the market. In the past, it was necessary to test subsystem and system performance at numerous set points covering the expected domain of operation of the system. For large, complex systems such a testing program can be prohibitively expensive, if not outright impossible, with available finite resources.

The current approach seeks to replace some or much of the experimentation with (cheaper) simulation results that have been validated with experimental results at selected set points—but to do this with confidence one must know how good the predictions are at the set points selected for validation and also how good the predictions are in the interpolated or extrapolated regions of the operating domain for which there are no experimental results. This has led to the emergence of the field called verification and validation (V&V) of simulations (e.g., models, codes).

Verification refers to application of approaches to determine that the algorithms solve the equations in the model correctly and to estimate the numerical uncertainty if the equations are discretized as, for example, in the finite-difference, finite-element, and finite-volume approaches used in computational mechanics. It addresses the question of whether the equations are solved correctly but does not address the question of how well the equations represent the real world. *Validation* is the process of determining the degree to which a model is an accurate representation of the real world. It addresses the question of how good the predictions are by comparison to experimental results (the real world) at the validation set points.

This chapter presents the verification and validation (V&V) methodology from ASME V&V20—2009 [1]: *Standard for Verification and Validation in Computational Fluid Dynamics and Heat Transfer*. Despite the apparent limitation in the title, these V&V procedures can be applied in all areas of computational engineering and science—the examples used in the document are from fluid dynamics and heat transfer. In V&V, the ultimate goal of engineering and scientific interest is validation, the process of determining the degree to which a model is an accurate representation of the real world from the perspective of the intended uses of the model. However, validation must be preceded by code verification and solution verification. Code verification establishes that the code accurately solves the conceptual model incorporated in the code, that is, that the code is free of mistakes. Solution verification estimates the numerical accuracy of a particular calculation. Both code and solution verification are discussed in detail in Ref. 1 and will be addressed only briefly in this chapter.

The estimation of a range within which the simulation modeling error lies is a primary objective of the validation process and is accomplished by comparing a simulation result (solution) with an appropriate experimental result (data) for specified validation variables at a specified set of conditions. There can be no validation without experimental data with which to compare the result of the simulation.

The V&V20 approach quantifies the degree of accuracy inferred from the comparison of solution and data for a specified variable at a specified validation point. The approach, initially proposed in Ref. 2, uses the concepts from experimental uncertainty analysis described in the previous chapters to consider the errors and uncertainties in both the solution and the data. The specific scope of the approach is the quantification of the degree of accuracy of simulation of specified validation variables at a specified validation point for cases in which the conditions of the actual experiment are simulated.

The determination of the degree of accuracy of a simulation result at a set point other than a validation points i.e., in interpolated or extrapolated regions of the operating domain, is still an unresolved research area and is not considered here.

6-2 ERRORS AND UNCERTAINTIES

The concepts of errors and uncertainties traditionally used in experimental uncertainty analysis [3, 4] were applied in Ref. 2 to the value of a solution variable from a simulation as well as a measured value of the variable from an experiment, and then the TSM propagation was applied to the comparison of the simulation and experimental results to estimate a range within which the simulation modeling error falls. Some other V&V references [5, 6] use different definitions of errors and uncertainties.

Following the previous discussion in this book, then, the validation approach presented in this chapter considers an error δ as a quantity that has a particular sign and magnitude; it is assumed that each error whose sign and magnitude are known has been removed by correction; and any remaining error is thus of unknown sign and magnitude. An uncertainty u is thus estimated with the idea that the range $\pm u$ contains the value of δ at some degree of confidence.

6-3 VALIDATION NOMENCLATURE

In the validation process, a simulation result (solution) is compared with an experimental result (data) for specified validation variables at a specified set of conditions (validation point). As an example, consider the case where we want to validate a computational fluid dynamics (CFD) code for the prediction of pressure drop and friction factor in a rough-walled pipe. An example of an experimental setup that might be used for the validation experiment (shown schematically in Figure 6.1) is fully developed flow in a rough-walled pipe with pressure drop ΔP (measured over some axial distance L) and friction factor f being the validation variables of interest. These validation variables are predicted by simulation and determined by experiment for a specified set of conditions (e.g., Reynolds number, pipe cross-sectional geometry, wall roughness, and Newtonian liquid).

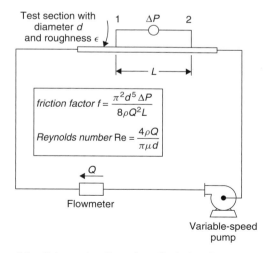

Figure 6.1 Schematic of rough-walled pipe flow experiment.

A classic experiment of this kind was executed and reported by Nikuradse [7], who investigated the effect of wall roughness on fully developed flow in circular pipes that had been roughened by gluing a specific size of sand onto the interior wall of a given pipe test section. The results of the experimental investigation were reported with friction factor f as a function of Reynolds number Re and *relative sand-grain roughness* ϵ/d.

Friction factor is defined in terms of measured variables as

$$f = \frac{\pi^2 d^5 (\Delta P)}{8\rho Q^2 L} \tag{6.1}$$

and Reynolds number as

$$\text{Re} = \frac{4\rho Q}{\pi \mu d} \tag{6.2}$$

and the third nondimensional group is ϵ/d, where ϵ in this example is taken as the *equivalent sand-grain roughness* (which is a single-length-scale descriptor of the roughness). In a more sophisticated treatment, the roughness might be characterized using more descriptors, that is, by height ϵ_1, spacing in the circumferential direction ϵ_2, spacing in the axial direction ϵ_3, shape factor s, and so on.

In the validation experiment, values for the following variables are found by measurement or from reference sources (property tables, dimensioned drawings):

- d, pipe diameter
- ϵ, roughness "size"
- L, distance between pressure taps
- ΔP, directly measured pressure drop over distance L
- Q, volumetric flow rate of fluid
- ρ, density of fluid
- μ, dynamic viscosity of fluid
- P_1, static pressure at position 1

The experimental value of f is calculated from the data reduction equation (6.1). Note that the specific validation point (Re, ϵ/d) is also calculated using the experimental values.

In the simulation, the experimentally determined values of d, ϵ, L, Q, ρ, μ and P_1 are inputs to the model. A value of ΔP is predicted by the model, and a predicted value of friction factor f is then calculated using its definition, Eq. (6.1).

The nomenclature used in the validation appoach is shown in Figure 6.2 using the pipe flow example with pressure drop ΔP as the validation variable and for a given validation condition (the experimentally determined values of Re and ϵ/d).

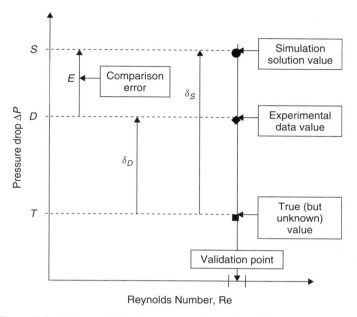

Figure 6.2   Schematic showing nomenclature for validation approach.

6-4 VALIDATION APPROACH

As shown in Figure 6.2, we will denote the predicted value of pressure drop ΔP from the simulation solution as S, the value of ΔP determined from experimental data as D, and the true (but unknown) value of pressure drop ΔP as T. (Obviously, the relative magnitudes of S, D, and T will differ from case to case and will not necessarily be in the order shown in the figure.) The validation comparison error E is defined as

$$E = S - D \tag{6.3}$$

The error in the solution value S is the difference between S and the true value T,

$$\delta_S = S - T \tag{6.4}$$

and similarly the error in the experimental value D is

$$\delta_D = D - T \tag{6.5}$$

Using Equations (6.3) to (6.5), E can be expressed as

$$E = S - D = (T + \delta_S) - (T + \delta_D) = \delta_S - \delta_D \tag{6.6}$$

The validation comparison error E is thus the combination of all of the errors in the simulation result and the experimental result, and its sign and magnitude are known once the validation comparison is made. All errors in S can be assigned to one of three categories [2]:

- the error δ_{model} due to modeling assumptions and approximations;
- the error δ_{num} due to the numerical solution of the equations, and
- the error δ_{input} in the simulation result due to errors in the simulation input parameters (d, ϵ, L, Q, ρ, μ, and P_1, for example).

Thus

$$\delta_S = \delta_{\text{model}} + \delta_{\text{num}} + \delta_{\text{input}} \qquad (6.7)$$

As we will discuss, there are ways to estimate the effects of δ_{num} and δ_{input}, but there are no ways to independently observe or calculate the effects of δ_{model}. The objective of a validation exercise is to estimate δ_{model} to within an uncertainty range.

Combining Eqs. (6.6) and (6.7), the comparison error can then be written as

$$E = \delta_{\text{model}} + \delta_{\text{num}} + \delta_{\text{input}} - \delta_D \qquad (6.8)$$

This approach is shown schematically in Figure 6.3, where the sources of error are shown in the ovals.

Rearranging Eq. (6.8) to isolate δ_{model} gives

$$\delta_{\text{model}} = E - (\delta_{\text{num}} + \delta_{\text{input}} - \delta_D) \qquad (6.9)$$

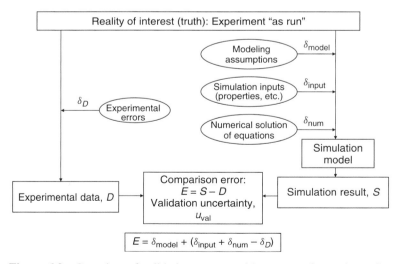

Figure 6.3 Overview of validation process with sources of error in ovals.

Consider the terms on the RHS of the equation. Once S and D are determined, the sign and magnitude of E are known from Eq. (6.3). However, the signs and magnitudes of δ_{num}, δ_{input}, and δ_D are unknown. The standard uncertainties corresponding to these errors are u_{num}, u_{input}, and u_D (where u_{num}, for instance, is the estimate of the standard deviation of the parent distribution from which δ_{num} is a single realization).

Following Ref. 2, a *validation uncertainty* u_{val} can be defined as an estimate of the standard deviation of the parent population of the combination of errors $(\delta_{num} + \delta_{input} - \delta_D)$. If the three errors are independent, then

$$u_{val} = \sqrt{u_{num}^2 + u_{input}^2 + u_D^2} \qquad (6.10)$$

(Cases in which S and D contain shared variables and/or errors from identical sources—and thus δ_{input} and δ_D are not independent—are discussed in the following sections.)

Considering the relationship shown in Eq. (6.9),

$$(E \pm u_{val})$$

then defines an interval within which δ_{model} falls (with some unspecified degree of confidence). To obtain an estimate of u_{val}, an estimate of u_{num} must be made; estimates must be made of the uncertainties in all input parameters that contribute to u_{input}; and estimates of the uncertainties in the experiment that contribute to u_D must be made.

The estimation of u_{val} is at the core of this methodology since knowledge of E and u_{val} allows determination of an interval within which the modeling error δ_{model} falls. Two uncertainty propagation approaches to estimating u_{val} are presented—the Taylor Series Method (TSM) and the Monte Carlo Method (MCM). These methods are demonstrated for the pipe flow example for cases in which (1) the directly measured pressure drop ΔP is the experiment validation variable, (2) the friction factor calculated from the directly measured pressure drop ΔP is the experiment validation variable, and (3) the pressure drop ΔP is the experiment validation variable and is calculated (rather than being measured directly) from measured values of P_1 and P_2 that contain an identical error source. A fourth case is discussed using a heat transfer example in which the experimentally determined value of wall heat flux (the validation variable) is inferred from temperature–time histories using a data reduction equation that is itself a model.

Note that once D and S have been determined, their values are always different by the same amount from the true value T. That is, all errors affecting D and S have become "fossilized" and δ_D, δ_{input}, δ_{num}, and δ_{model} are all *systematic* errors. This means that the uncertainties to be estimated (u_{input}, u_{num}, and u_D) are systematic uncertainties.

The estimates of the uncertainties in all input parameters that contribute to u_{input} and the estimates of the uncertainties in the experiment that contribute to u_D are made using the techniques described in previous chapters. Details of the verification procedures used to estimate u_{num} are beyond the scope of this text—a brief overview is given in the following section.

6-5 CODE AND SOLUTION VERIFICATION

Prior to estimating u_{num} it is necessary to verify the code itself, that is, to determine that the code is free of mistakes (code verification). Solution verification is then the process to estimate u_{num}, which is the uncertainty estimate required for the validation process. Code and solution verification are mathematical activities, with no consideration of the agreement of the numerical model results with physical data from experiments (which is the concern of validation). Code and solution verification are still active research areas as this is written, and for a detailed discussion the reader is referred to V&V20 [1], which covers verification methods that are specific to grid-based simulations, and also to Roache [8].

For such simulations, the recommended [1] approach for code verification is the use of the method of manufactured solutions (MMS). The MMS assumes a sufficiently complex solution form (e.g., hyperbolic tangent, tanh, or other transcendental function) so that all of the terms in the partial differential equations (PDEs) are exercised. The solution is input to the PDEs as a source term, and grid convergence tests are performed on the code not only to verify that it converges but also to ascertain at what rate it converges. The magnitude (and sign) of the error is directly computed from the difference between the numerical solution and the analytical solution.

Whereas grid refinement studies in the context of code verification provide an *evaluation* of error, grid refinement studies used in solution verification provide an *estimate* of error. The most widely used method to obtain such an error estimate is classical Richardson extrapolation (RE) [9]. Uncertainty estimates at a given degree of confidence can then be calculated by Roache's [8] grid convergence index (GCI). The GCI is an estimated 95% uncertainty obtained by multiplying the (generalized) RE error estimate by an empirically determined factor of safety, Fs. The Fs is intended to convert the best error estimate implicit in the definition of any ordered error estimate (like RE) into a 95% uncertainty estimate. The GCI, especially the least-squares versions pioneered by Eça and Hoekstra [10], is cited [1] as the most robust and tested method available for the prediction of numerical uncertainty as of this date.

6-6 ESTIMATION OF VALIDATION UNCERTAINTY u_{val}

Once an estimate of u_{num} has been made using the techniques in Section 6-5 and the uncertainty contributors to u_{input} and u_D have been made, u_{val} can be obtained by several approaches. The two approaches illustrated here are the TSM and the MCM. Both approaches are illustrated for four example cases that cover a wide range of V&V applications.

The first three cases considered are for flow in the rough-walled pipe discussed previously and shown schematically in Figure 6.1. In case 1, the experiment validation variable (ΔP) is directly measured; in case 2, the experiment validation variable (friction factor) is a result defined by a data reduction equation that combines variables measured in the experiment (where the pressure drop ΔP is directly measured); and in case 3, the experiment validation variable (ΔP) is the result defined by a data reduction equation ($\Delta P = P_1 - P_2$) that combines variables measured in the experiment *and* the measurements of P_1 and P_2 share identical error sources. In all of these cases, specification of the validation condition (set point) requires experimental determination of the values of Reynolds number ($4\rho Q/\pi\mu d$) and the relative roughness ϵ/d, and since the simulation is performed for actual experimental conditions, the values of the variables P_1, d, L, ϵ, ρ, μ and Q (at a minimum) from the experiment will be inputs to the simulation. The systematic errors in these inputs are assumed to be uncorrelated for all cases, with the exception of P_1 and P_2 for case 3.

Case 4 considers a combustion flow with duct wall heat flux q at a given location being the validation variable (Figure 6.4). The experimental q is inferred by applying an inverse conduction heat transfer method [11] to temperature–time measurements at the outside duct wall using a data reduction equation that is itself a model. The predicted q is from a simulation using a turbulent chemically reacting flow code to model the flow through the duct.

6-6.1 Estimating u_{val} When Experimental Value D of Validation Variable Is Directly Measured (Case 1)

This case is one in which the experimental value D of the validation variable is directly measured. A key feature of such cases is that D and S have no shared variables, which leads to a straightforward evaluation of u_{input} and u_D. The analysis is more complex in cases for which D and S have shared variables, as shown in the following sections for cases 2 and 3.

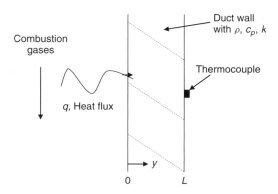

Figure 6.4 Schematic for V&V example case of combustion gas flow through a duct.

Using the rough-walled pipe flow experiment shown in Figure 6.1 as an example, consider a case in which the experiment validation variable is defined as the directly measured pressure difference ΔP. Then

$$S = \Delta P_S = P_{2,S} - P_{1,D} \tag{6.11}$$

$$D = \Delta P_D \tag{6.12}$$

$$E = S - D = \Delta P_S - \Delta P_D \tag{6.13}$$

The functional dependence of the simulation result is represented by

$$\Delta P_S = \Delta P_S(P_1, d, L, \epsilon, \rho, \mu, Q) \tag{6.14}$$

(where the simulation models the conditions of the experiment, so that the values of P_1, d, L, ϵ, ρ, μ, and Q from the experiment are used as inputs to the simulation). The expression for the comparison error is then

$$E = \Delta P_S(P_1, d, L, \epsilon, \rho, \mu, Q) - \Delta P_D \tag{6.15}$$

TSM Approach. Since the experiment validation variable ΔP_D is directly measured, the experiment and the simulation share no variables and the assumption of effectively independent errors δ_{input} and δ_D is reasonable. The expression for u_{val} is, from Eq (6.10),

$$u_{\text{val}}^2 = u_{\text{num}}^2 + u_{\text{input}}^2 + u_{\Delta P_D}^2$$

with u_{input} given by the TSM with correlation terms equal to zero,

$$u_{\text{input}}^2 = \sum_{i=1}^{J} \left(\frac{\partial \Delta P_s}{\partial x_i} u_{x_i} \right)^2 \tag{6.16}$$

which for this particular case yields

$$u_{\text{input}}^2 = \left(\frac{\partial \Delta P_s}{\partial P_1} \right)^2 u_{P_1}^2 + \left(\frac{\partial \Delta P_s}{\partial d} \right)^2 u_d^2 + \left(\frac{\partial \Delta P_s}{\partial L} \right)^2 u_L^2 + \left(\frac{\partial \Delta P_s}{\partial \epsilon} \right)^2 u_\epsilon^2$$

$$+ \left(\frac{\partial \Delta P_s}{\partial \rho} \right)^2 u_\rho^2 + \left(\frac{\partial \Delta P_s}{\partial \mu} \right)^2 u_\mu^2 + \left(\frac{\partial \Delta P_s}{\partial Q} \right)^2 u_Q^2 \tag{6.17}$$

These derivatives for the simulation are probably obtained using numerical methods as discussed earlier in Section 3-1.5.

Uncertainty exists in the validation condition set point due to uncertainties in the parameters defining the set point. Applying the TSM approach to Eq. (6.2) and to ϵ/d leads to

$$u_{\text{Re}}^2 = \left(\frac{\partial \text{Re}}{\partial \rho}\right)^2 u_\rho^2 + \left(\frac{\partial \text{Re}}{\partial Q}\right)^2 u_Q^2 + \left(\frac{\partial \text{Re}}{\partial \mu}\right)^2 u_\mu^2 + \left(\frac{\partial \text{Re}}{\partial d}\right)^2 u_d^2 \qquad (6.18)$$

$$u_{\epsilon/d}^2 = \left(\frac{\partial (\epsilon/d)}{\partial \epsilon}\right)^2 u_\epsilon^2 + \left(\frac{\partial (\epsilon/d)}{\partial d}\right)^2 u_d^2 \qquad (6.19)$$

These derivatives can be evaluated analytically due to the simple forms of Re and ϵ/d.

A graphical summary of the procedures used to evaluate u_{val} using the TSM propagation approach is illustrated in Figure 6.5.

MCM Approach. Figure 6.6 illustrates the Monte Carlo approach for this case. In contrast to the TSM approach, the MCM requires that probability distributions for the errors in the experiment and the errors in the input parameters be assumed. The standard uncertainties u are generally taken to be the standard deviations of the assumed distributions. For a given "run" of the experiment and simulation, an error is taken from each of these distributions; the experimental result D_i and the simulation result S_i are calculated; and the corresponding validation comparison error E_i and validation point $(\text{Re}_i, (\epsilon/d)_i)$ evaluated. This process is repeated M times, and the resulting means and standard deviations of the M values of E_i, Re_i, and $(\epsilon/d)_i)$ evaluated.

Since the simulation is run M times at the validation point (or actually, in the perturbed vicinity of the validation point), δ_{model} is constant to first order. Note that since each S_i includes (essentially) the same δ_{num}, the effect of δ_{num} is not observed in the variability of the distribution of the M values of S_i or E_i. The effect of the numerical uncertainty is accounted for when u_{num} is included in the calculation of u_{val}.

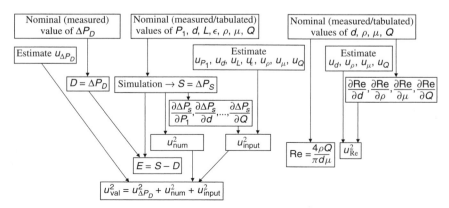

Figure 6.5 Case 1: TSM approach for estimating u_{val} when experiment validation variable (ΔP) is directly measured.

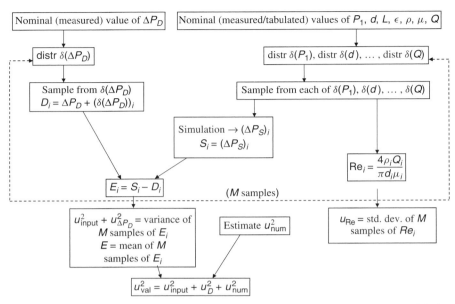

Figure 6.6 Case 1: MCM approach for estimating u_{val} when experiment validation variable (ΔP) is directly measured.

The standard deviation of the sample of M values of E_i is an estimate of $\sqrt{u_D^2 + u_{\text{input}}^2}$. The number of iterations M can be minimized using the techniques discussed in V&V20 [1].

6-6.2 Estimating u_{val} When Experimental Value D of Validation Variable Is Determined from Data Reduction Equation (Cases 2 and 3)

When the experiment validation variable is not directly measured but is determined from a data reduction equation using other measured variables, the estimation of u_{input} and u_D (and subsequently u_{val}) becomes more complex. Example cases 2 and 3 illustrate the application of the validation approach in such circumstances. The most general form of the TSM approach as it applies to these cases is presented first, with the form for each of the two specific cases then presented in the sections following.

Consider the general situation in which the experiment and simulation validation variables are results determined from data reduction equations each containing some of the J variables x_i where some of the measured variables may share identical error sources. The general form of the equation for the comparison error is then (where S and D are shown to be functions of all of the J variables)

$$E = S(x_1, x_2, \ldots, x_J) - D(x_1, x_2, \ldots, x_J) = \delta_{\text{model}} + \delta_{\text{num}} + \delta_{\text{input}} - \delta_D$$

$$(6.20)$$

In this instance, δ_{input} and δ_D cannot reasonably be assumed to be independent since S and D share a dependence on the same measured variables. Application of the TSM approach to obtain an expression for u_{val} yields

$$
u_{\mathrm{val}}^2 = \left[\left(\frac{\partial S}{\partial x_1}\right) - \left(\frac{\partial D}{\partial x_1}\right)\right]^2 u_{x_1}^2 + \left[\left(\frac{\partial S}{\partial x_2}\right) - \left(\frac{\partial D}{\partial x_2}\right)\right]^2 u_{x_2}^2 + \cdots
$$
$$
+ \left[\left(\frac{\partial S}{\partial x_j}\right) - \left(\frac{\partial D}{\partial x_j}\right)\right]^2 u_{x_j}^2 + 2\left[\left(\frac{\partial S}{\partial x_1}\right) - \left(\frac{\partial D}{\partial x_1}\right)\right]
$$
$$
\times \left[\left(\frac{\partial S}{\partial x_2}\right) - \left(\frac{\partial D}{\partial x_2}\right)\right] u_{x_1 x_2} + \cdots + u_{\mathrm{num}}^2 \tag{6.21}
$$

where there is a covariance term containing a $u_{x_1 x_2}$ factor for each pair of x variables that share identical error sources. When S or D have no dependence on a variable x_i, those derivatives will be zero. There is no explicit expression for u_{input}^2 as its components combine implicitly with components of u_D^2. Equation (6.21) can be expressed in a form analogous to Eq. (6.10) as

$$
u_{\mathrm{val}}^2 = u_{\mathrm{num}}^2 + u_{\mathrm{input}+D}^2 \tag{6.22}
$$

where

$$
u_{\mathrm{input}+D}^2 = \left[\left(\frac{\partial S}{\partial x_1}\right) - \left(\frac{\partial D}{\partial x_1}\right)\right]^2 u_{x_1}^2 + \left[\left(\frac{\partial S}{\partial x_2}\right) - \left(\frac{\partial D}{\partial x_2}\right)\right]^2 u_{x_2}^2 + \cdots
$$
$$
+ \left[\left(\frac{\partial S}{\partial x_J}\right) - \left(\frac{\partial D}{\partial x_J}\right)\right]^2 u_{x_J}^2 + 2\left[\left(\frac{\partial S}{\partial x_1}\right) - \left(\frac{\partial D}{\partial x_1}\right)\right]
$$
$$
\times \left[\left(\frac{\partial S}{\partial x_2}\right) - \left(\frac{\partial D}{\partial x_2}\right)\right] u_{x_1 x_2} + \cdots + 2\left[\left(\frac{\partial S}{\partial x_{J-1}}\right) - \left(\frac{\partial D}{\partial x_{J-1}}\right)\right]
$$
$$
\times \left[\left(\frac{\partial S}{\partial x_J}\right) - \left(\frac{\partial D}{\partial x_J}\right)\right] u_{x_{J-1} x_J} \tag{6.23}
$$

6-6.2.1 No Measured Variables Share Identical Error Sources (Case 2)

6-6.2.1 No Measured Variables Share Identical Error Sources (Case 2)   Again using the rough-walled pipe flow experiment as an example, consider now a case in which the experiment validation variable of interest is the friction factor f defined by Eq. (6.1):

$$
f = \frac{\pi^2 d^5 \, \Delta P}{8\rho Q^2 L}
$$

and the pressure difference ΔP is directly measured. It is important to note several points. First, the friction factor f is not directly measured—it is an experimental result determined from measured variables and others whose values are found from reference sources (e.g., the properties). Second, since Eq. (6.1) is the *definition* of friction factor, there is no modeling error incurred when it is used as contrasted with the situation to be discussed in case 4. In this example, it is assumed that the simulation predicts the pressure drop ΔP_S and then calculates f_S. Since for this example there are no error sources shared by the different variables used to calculate D and S, all covariance terms in the uncertainty propagation equation are zero.

The comparison error expression is

$$E = S - D = f_S - f_D \tag{6.24}$$

where, using Eq. (6.1),

$$f_S = \frac{\pi^2 d^5 \, \Delta P_S}{8 \rho Q^2 L} \tag{6.25}$$

and

$$f_D = \frac{\pi^2 d^5 \, \Delta P_D}{8 \rho Q^2 L} \tag{6.26}$$

Eq. (6.24) can be written as

$$E = f_S - f_D = \frac{\pi^2 d^5 \, \Delta P_S(P_1, d, L, \epsilon, \rho, \mu, Q)}{8 \rho Q^2 L} - \frac{\pi^2 d^5 \, \Delta P_D}{8 \rho Q^2 L} \tag{6.27}$$

TSM Approach. For this case u_{val} is given by Eqs. (6.22) and (6.23) where $u_{\text{input}+D}$ is expressed as

$$
\begin{aligned}
(u_{\text{input}+D}^2) = & \left[\left(\frac{\partial f_S}{\partial d} \right) - \left(\frac{\partial f_D}{\partial d} \right) \right]^2 u_d^2 + \left[\left(\frac{\partial f_S}{\partial L} \right) - \left(\frac{\partial f_D}{\partial L} \right) \right]^2 u_L^2 \\
& + \left[\left(\frac{\partial f_S}{\partial \rho} \right) - \left(\frac{\partial f_D}{\partial \rho} \right) \right]^2 u_\rho^2 + \left[\left(\frac{\partial f_S}{\partial Q} \right) - \left(\frac{\partial f_D}{\partial Q} \right) \right]^2 u_Q^2 \\
& + \left(\frac{\partial f_S}{\partial P_1} \right)^2 u_{P_1}^2 + \left(\frac{\partial f_S}{\partial \epsilon} \right)^2 u_\epsilon^2 + \left(\frac{\partial f_S}{\partial \mu} \right)^2 u_\mu^2 + \left(\frac{\partial f_D}{\partial \Delta P_D} \right)^2 u_{\Delta P_D}^2
\end{aligned} \tag{6.28}
$$

Equations (6.18) and (6.19) are used to evaluate the uncertainty in the set point (Re, ϵ/d). Figure 6.7 illustrates the application of the TSM approach to this example.

MCM Approach. As in the previous case, probability distributions for the errors in the experiment and the errors in the input parameters are assumed; the standard uncertainties u are taken to be the standard deviations of the assumed distributions; and the standard deviation of the sample of M values of E_i is an estimate of $\sqrt{u_{\text{input}+D}^2}$. The MCM approach is illustrated in Figure 6.8.

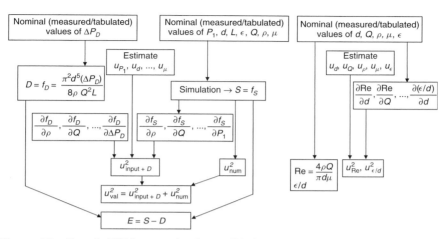

Figure 6.7 Case 2: TSM approach when validation result is defined by data reduction equation that combines variables measured in experiment.

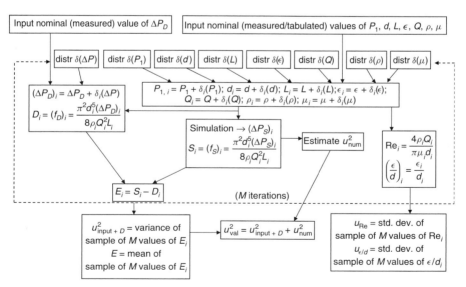

Figure 6.8 Case 2: MCM approach when validation result is defined by data reduction equation that combines variables measured in experiment.

6-6.2.2 Measured Variables Share Identical Error Sources (Case 3) Again
using the rough-walled pipe flow experiment as an example, consider now
the case in which the validation variable of interest is the pressure drop
$\Delta P = P_1 - P_2$ and the measurements of P_1 and P_2 share identical error sources.
Then

$$S = \Delta P_S = P_{1,D} - P_{2,S} \tag{6.29}$$

$$D = \Delta P_D = P_{1,D} - P_{2,D} \tag{6.30}$$

$$E = S - D = \Delta P_S - \Delta P_D = (P_{1,D} - P_{2,S}) - (P_{1,D} - P_{2,D}) \tag{6.31}$$

where the subscript D on P_1 is included to emphasize that P_1 is an experimentally
measured value. The subscript D on P_1 is dropped in the equations that follow.

The functional dependence of the simulation result is represented as

$$\Delta P_S = P_1 - P_{2,S}(P_1, d, L, \epsilon, \rho, \mu, Q) \tag{6.32}$$

(where the simulation models the conditions of the experiment, so that the values
of P_1, d, L, ϵ, ρ, μ, and Q from the experiment are used as inputs to the
simulation). The expression for the comparison error is then

$$E = P_1 - P_{2,S}(P_1, d, L, \epsilon, \rho, \mu, Q) - P_1 + P_{2,D}$$
$$= -P_{2,S}(P_1, d, L, \epsilon, \rho, \mu, Q) + P_{2,D} \tag{6.33}$$

TSM Approach. For this case u_{val} is given by Eqs. (6.22) and (6.23) where
$u_{input+D}$ is expressed as

$$u_{input+D}^2 = \left(\frac{\partial P_{2,S}}{\partial P_1}\right)^2 u_{P_1}^2 + \left(\frac{\partial P_{2,S}}{\partial d}\right)^2 u_d^2 + \left(\frac{\partial P_{2,S}}{\partial L}\right)^2 u_L^2 + \left(\frac{\partial P_{2,S}}{\partial \epsilon}\right)^2 u_\epsilon^2$$
$$+ \left(\frac{\partial P_{2,S}}{\partial \rho}\right)^2 u_\rho^2 + \left(\frac{\partial P_{2,S}}{\partial \mu}\right)^2 u_\mu^2 + \left(\frac{\partial P_{2,S}}{\partial Q}\right)^2 u_Q^2 + u_{P_{2,D}}^2$$
$$+ 2\left[\left(\frac{-\partial P_{2,S}}{\partial P_1}\right) + \left(\frac{\partial P_{2,D}}{\partial P_1}\right)\right]\left[\left(\frac{-\partial P_{2,S}}{\partial P_{2,D}}\right) + \left(\frac{\partial P_{2,D}}{\partial P_{2,D}}\right)\right] u_{P_1 P_{2,D}} \tag{6.34}$$

where the final term in the equation is the covariance term that takes into account
the fact that the measured values of P_1 and $P_{2,D}$ share an identical error from
the same source. Substituting into the final term of Eq. (6.34) for the derivatives
that are zero and unity, the equation becomes

$$u_{input+D}^2 = \left(\frac{\partial P_{2,S}}{\partial P_1}\right)^2 u_{P_1}^2 + \left(\frac{\partial P_{2,S}}{\partial d}\right)^2 u_d^2 + \left(\frac{\partial P_{2,S}}{\partial L}\right)^2 u_L^2 + \left(\frac{\partial P_{2,S}}{\partial \epsilon}\right)^2 u_\epsilon^2$$

$$+ \left(\frac{\partial P_{2,S}}{\partial \rho}\right)^2 u_\rho^2 + \left(\frac{\partial P_{2,S}}{\partial \mu}\right)^2 u_\mu^2 + \left(\frac{\partial P_{2,S}}{\partial Q}\right)^2 u_Q^2 + u_{P_{2,D}}^2 \qquad (6.35)$$

$$+ 2\left(\frac{-\partial P_{2,S}}{\partial P_1}\right) u_{P_1 P_{2,D}}$$

Equations (6.18) and (6.19) are used to evaluate the uncertainty in the set point (Re, ϵ/d). Figure 6.9 illustrates the application of the TSM approach to this example.

MCM Approach. As in the previous cases, probability distributions for the errors in the experiment and the errors in the input parameters are assumed; the standard uncertainties u are taken to be the standard deviations of the assumed distributions; and the variance of the sample of M values of E_i is taken as the estimate of $u_{\mathrm{input}+D}^2$. Note from Figure 6.10 that on a given iteration a single pressure measurement error is drawn from distr $\delta(P)$ and assigned to both $P_{1,i}$ and $P_{2,i}$.

6-6.3 Estimating u_{val} When Experimental Value D of Validation Variable Is Determined from Data Reduction Equation That Itself Is a Model (Case 4)

Consider the case of combustion gases flowing through a duct, with the heat flux q incident on a particular area of the duct wall the validation variable of interest. The situation is shown schematically in Figure 6.4. The simulation result q_S is predicted using a code which models a turbulent chemically reacting flow

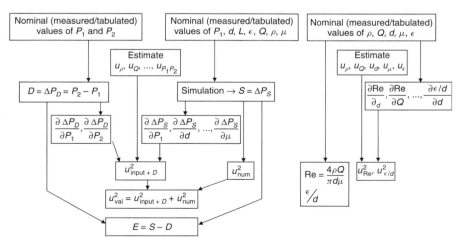

Figure 6.9 Case 3: TSM approach when validation result is defined by data reduction equation that combines variables measured in experiment that share identical error source.

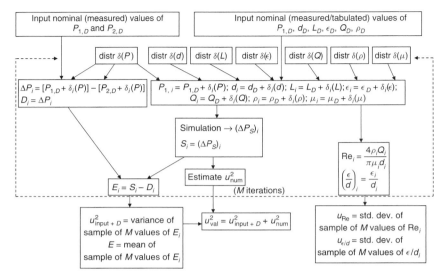

Figure 6.10 Case 3: MCM approach when the validation result is defined by a data reduction equation that combines variables measured in the experiment that share an identical error source.

at the conditions of the experiment. Inputs would be, for example, geometry and propellant and oxidizer flow rates. The chemical equilibrium code that calculates properties of the combustion gases might be considered to be a part of the simulation model (similar to the treatment of turbulence models in a CFD analysis).

The experimental heat flux is determined using two approaches. The first is direct measurement of the heat flux using a calibrated calorimeter mounted in the wall. For this approach the analysis to determine u_{val} is analogous to that in the pipe flow example in which the experiment validation variable of interest is the directly measured pressure drop (Section 6-6.1). The second approach, discussed in detail below, is to measure the temperature of the back wall ($y = L$) of the duct as a function of time. The measured temperature (T)–time (t) history is then used in an inverse conduction model [11] to infer the incident heat flux at $y = 0$ (Figure 6.4). The model might assume one-dimensional conduction, constant or variable wall properties, incident heat flux constant with time, and so on. In this approach, the experimental result q_D now contains errors from categories analogous to those in the simulation, that is, the error due to assumptions and approximations in the data reduction model, which is denoted $\delta_{D,\mathrm{model}}$; the error in the model output due to the errors in the inputs (measured and from reference sources), which is denoted $\delta_{D,\mathrm{input}}$; and the error due to the numerical solution of the model, which is denoted $\delta_{D,\mathrm{num}}$.

The validiation comparison error in this case is given by

$$E = S - D = q_S - q_D$$

$$= \delta_{S,\mathrm{model}} + \delta_{S,\mathrm{input}} + \delta_{S,\mathrm{num}} - \delta_{D,\mathrm{model}} - \delta_{D,\mathrm{input}} - \delta_{D,\mathrm{num}} \quad (6.36)$$

Consider first the case in which $\delta_{D,\text{model}}$ *is not* (or cannot be) estimated with an uncertainty. Then the two modeling errors are not distinguishable individually and a total modeling error is given by

$$\delta_{\text{model,total}} = (\delta_{S,\text{model}} - \delta_{D,\text{model}}) = E - (\delta_{S,\text{input}} + \delta_{S,\text{num}} - \delta_{D,\text{input}} - \delta_{D,\text{num}})$$
$$(6.37)$$

Now u_{val} is defined as the standard uncertainty corresponding to the standard deviation of the parent population of the combination of $\delta_{S,\text{input}} + \delta_{S,\text{num}} - \delta_{D,\text{input}} - \delta_{D,\text{num}}$.

The functional relationships for q_S and q_D are given by

$$q_S = q_S(x_1, x_2, \ldots, x_J) \qquad (6.38)$$

where the J different x_i are the inputs to the simulation model, and

$$q_D = q_D(\rho, c_p, k, L, T, t) \qquad (6.39)$$

Realizing that the simulation is of the flow field and the experimental model is of the duct wall, the expressions for the results q_S and q_D do not contain shared variables, as in cases 2 and 3 discussed in the previous sections. In this case, the additional errors $\delta_{D,\text{input}}$ and $\delta_{D,\text{num}}$ must be considered in the estimation of u_{val}.

TSM Approach. The TSM approach in this case yields

$$u_{\text{val}}^2 = \left(\frac{\partial q_S}{\partial x_1}\right)^2 u_{x_1}^2 + \cdots + \left(\frac{\partial q_S}{\partial x_J}\right)^2 u_{x_J}^2 + u_{S,\text{num}}^2 + \left(\frac{\partial q_D}{\partial \rho}\right)^2 u_\rho^2 + \left(\frac{\partial q_D}{\partial c_p}\right)^2 u_{c_p}^2$$
$$+ \left(\frac{\partial q_D}{\partial k}\right)^2 u_k^2 + \left(\frac{\partial q_D}{\partial L}\right)^2 u_L^2 + \left(\frac{\partial q_D}{\partial T}\right)^2 u_T^2 + \left(\frac{\partial q_D}{\partial t}\right)^2 u_t^2 + u_{D,\text{num}}^2$$
$$(6.40)$$

Defining

$$u_{S,\text{input}}^2 = \left(\frac{\partial q_S}{\partial x_1}\right)^2 u_{x_1}^2 + \cdots + \left(\frac{\partial q_S}{\partial x_J}\right)^2 u_{x_J}^2 \qquad (6.41)$$

and

$$u_{D,\text{input}}^2 = \left(\frac{\partial q_D}{\partial \rho}\right)^2 u_\rho^2 + \left(\frac{\partial q_D}{\partial c_p}\right)^2 u_{c_p}^2 + \left(\frac{\partial q_D}{\partial k}\right)^2 u_k^2 + \left(\frac{\partial q_D}{\partial L}\right)^2 u_L^2$$
$$+ \left(\frac{\partial q_D}{\partial T}\right)^2 u_T^2 + \left(\frac{\partial q_D}{\partial t}\right)^2 u_t^2 \qquad (6.42)$$

the expression for u_{val} becomes

$$u_{\text{val}}^2 = u_{S,\text{input}}^2 + u_{S,\text{num}}^2 + u_{D,\text{input}}^2 + u_{D,\text{num}}^2 \qquad (6.43)$$

Now consider the case in which $\delta_{D,\text{model}}$ *is* estimated with an uncertainty. Then the simulation modeling error is given by

$$\delta_{S,\text{model}} = E - (\delta_{S,\text{input}} + \delta_{S,\text{num}} - \delta_{D,\text{model}} - \delta_{D,\text{input}} - \delta_{D,\text{num}}) \qquad (6.44)$$

and u_{val} is defined as the standard uncertainty corresponding to the standard deviation of the parent population of the combination of $\delta_{S,\text{input}} + \delta_{S,\text{num}} - \delta_{D,\text{model}} - \delta_{D,\text{input}} - \delta_{D,\text{num}}$. In this case

$$u_{\text{val}}^2 = u_{S,\text{input}}^2 + u_{S,\text{num}}^2 + u_{D,\text{model}}^2 + u_{D,\text{input}}^2 + u_{D,\text{num}}^2 \qquad (6.45)$$

where $u_{S,\text{input}}^2$ is given by Eq. (6.41), $u_{D,\text{input}}^2$ is given by Eq. (6.42), and $u_{D,\text{model}}^2$ may well be estimated analytically with a parametric study using a detailed thermal model of the wall and sensor with a wide range of assumed conditions and perturbations. Figure 6.11 illustrates this case.

MCM Approach. As in the previous cases, probability distributions of the errors are assumed. For this case, all variables are input parameters, both the input parameters for the simulation and those for the experimental result; the standard uncertainties u are taken to be the standard deviations of the assumed distributions. The validation uncertainty is found as shown in Figure 6.12.

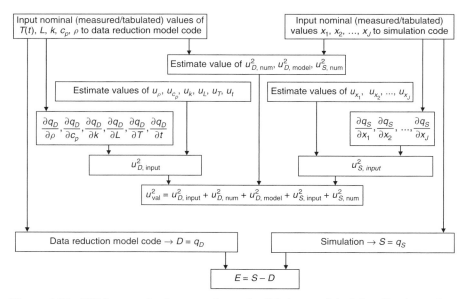

Figure 6.11 TSM approach when experimental validation result is defined by data reduction equation that itself is a model (Case 4).

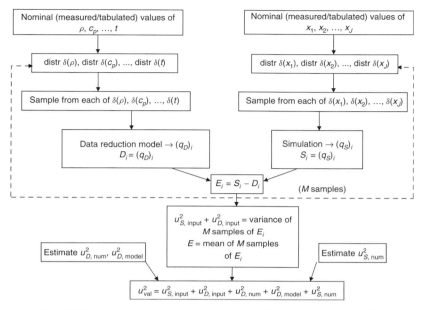

Figure 6.12 MCM approach when experimental validation result is defined by data reduction equation that itself is a model (case 4).

6-7 INTERPRETATION OF VALIDATION RESULTS USING E AND u_{val}

Recall Equation (6.9):

$$\delta_{\text{model}} = E - (\delta_{\text{input}} + \delta_{\text{num}} - \delta_D)$$

and that once a validation effort has determined values S and D of the validation variable, then the sign and magnitude of the validation comparison error $E = (S - D)$ are known. The validation uncertainty u_{val} is an estimate of the standard uncertainty corresponding to the standard deviation of the parent population of the combination of all errors ($\delta_{\text{input}} + \delta_{\text{num}} - \delta_D$) except the modeling error—examples for four specific cases have been discussed. Considering the relationship shown in Eq. (6.9), $E \pm u_{\text{val}}$ then defines an interval within which δ_{model} falls (with some unspecified degree of confidence). Thus E is an estimate of δ_{model} and u_{val} is the standard uncertainty of that estimate and can properly be designated $u_{\delta_{\text{model}}}$.

6-7.1 Interpretation with No Assumptions Made about Error Distributions

If one has only an estimate for the standard uncertainty $u_{\delta_{\text{model}}}$ of δ_{model} and not an estimate of the probability distribution associated with $\delta_{\text{input}} + \delta_{\text{num}} - \delta_D$, an interval within which the value of δ_{model} falls with a given probability cannot be estimated without further assumption. One can make the following statements, however:

1. If

$$|E| \gg u_{\text{val}} \tag{6.46}$$

then probably $\delta_{\text{model}} \approx E$.

2. If

$$|E| \leq u_{\text{val}} \tag{6.47}$$

then probably δ_{model} is of the same order as or less than $\delta_{\text{num}} + \delta_{\text{input}} - \delta_d$.

6-7.2 Interpretation with Assumptions Made about Error Distributions

In order to estimate an interval within which δ_{model} falls with a given degree of confidence, an assumption about the probability distribution of the combination of all errors except the modeling error must be made. This then allows the choice of a coverage factor k such that

$$U_{\%} = k_{\%} u \tag{6.48}$$

where U is the *expanded uncertainty* and one can say, for instance, that $E \pm U_{95}$ then defines an interval within which δ_{model} falls about 95 times out of 100 (i.e., with 95% confidence) when the coverage factor has been chosen for a level of confidence of 95%.

To obtain a perspective on the order of magnitude of k, consider the three parent error distributions used as examples in the ISO guide [3]—a uniform (rectangular) distribution with equal probability that δ lies at any value between $-A$ and $+A$ so that $\sigma = A/\sqrt{3}$; a triangular distribution symmetric about $\delta = 0$ with base from $-A$ to $+A$ so that $\sigma = A/\sqrt{6}$; and a Gaussian distribution with standard deviation σ. Choose a coverage factor k such that $\delta_{\text{num}} + \delta_{\text{input}} - \delta_d$ certainly (or almost certainly) falls within $\pm k(u_{\text{val}})$. If $\delta_{\text{num}} + \delta_{\text{input}} - \delta_d$ is from the uniform distribution, 100% of the population is covered for $k = 1.73$. If $\delta_{\text{num}} + \delta_{\text{input}} - \delta_d$ is from the triangular distribution, 100% of the population is covered for $k = 2.45$. If $\delta_{\text{num}} + \delta_{\text{input}} - \delta_d$ is from the Gaussian distribution, 99.7% of the population is covered for $k = 3.00$, 99.95% for $k = 3.5$, and 99.99% for $k = 4.0$. With these comparisons, one can conclude that for error distributions in the "family" of the three distributions considered, δ_{model} certainly (or almost certainly) falls within the interval $E \pm k(u_{\text{val}})$, where k is a small single-digit number of the order of 2 or 3.

In the case of the Monte Carlo approach, an expanded uncertainty with a given level of confidence can be determined using the techniques discussed in Chapter 3.

6-8 SOME PRACTICAL POINTS

Ideally, as a V&V program is initiated, those responsible for the simulations and those responsible for the experiments should be involved cooperatively in designing the V&V effort. The validation variables should be chosen and defined with care. Each measured variable has an inherent temporal and spatial resolution, and the experimental result that is determined from these measured variables should be compared with a predicted result that possesses the same spatial and temporal resolution. If this is not done, such conceptual errors must be identified and corrected or estimated in the initial stages of a V&V effort or substantial resources can be wasted and the entire effort compromised.

If, as demonstrated in the basic methodology discussed in this chapter, uncertainty contributions to u_{val} are considered that take into account all of the error sources in δ_{num}, δ_{input}, and δ_D, then δ_{model} includes only errors arising from modeling assumptions and approximations ("model form" errors). In practice, there are numerous gradations that can exist in the choices of which error sources are accounted for in δ_{input} and which error sources are defined as an inherent part of δ_{model}. The code used will often have more adjustable parameters or data inputs than the analyst may decide to use (e.g., for a commercial code). The decision of which parameters to include in the definition of the computational *model* (conceptually separate from the *code*) is somewhat arbitrary. Some (even all) of the parameters available may be considered fixed for the simulation. For example, an analyst may decide to treat parameters in a chemistry package as fixed ("hard wired") and therefore not to be considered in estimating u_{input}, even though these parameters had associated uncertainties. The point here is that the computational model which is being assessed consists of the code and a selected number of simulation inputs which are considered part of the model. *It is crucial in interpreting the results of a validation effort that which error sources are included in δ_{model} and which are accounted for in the estimation of u_{val} be defined precisely and unambiguously.*

REFERENCES

1. American Society of Mechanical Engineers (ASME), *Standard for Verification and Validation in Computational Fluid Dynamics and Heat Transfer*, V&V20—2009, ASME, New York, 2009.
2. Coleman, H. W., and Stern, F., "Uncertainties and CFD Code Validation," *Journal of Fluids Engineering*, Vol. 119, Dec. 1997, pp. 795–803. (Also "Discussion and Authors' Closure," Vol. 120, Sept. 1998, pp. 635–636.)
3. International Organization for Standardization (ISO), *Guide to the Expression of Uncertainty in Measurement* (corrected and reprinted 1995), ISO Geneva, 1993.
4. American Society for Mechanical Engineers (ASME), "Test Uncertainty", ASME PTC 19. 1–2005, ASME, New York, 2006.
5. American Institute of Aeronautics and Astronautics (AIAA), *Guide for Verification and Validation of Computational Fluid Dynamics Simulations*, AIAA G-077—1998, AIAA, New York, 1998.

6. American Society for Mechanical Engineers (ASME), *Guide for Verification and Validation in Computational Solid Mechanics*, ASME V&V 10–2006, ASME, New York, 2006.

7. Nikuradse, J., "Stromugsgestze in Rauhen Rohren," VDI Forschungsheft, No. 361, 1950 (English Translation, NACA TM 1292).

8. Roache, P. J., *Verification and Validation in Computational Science and Engineering*, Hermosa Publishers, Albuquerque, NM, 1998.

9. Richardson, L. F., "The Approximate Arithmetical Solution by Finite Differences of Physical Problems Involving Differential Equations, with an Application to the Stresses in a Masonry Dam," *Transactions of the Royal Society of London*, Ser. A, Vol. 210, 1910, pp. 307–357.

10. Eça, L., and Hoekstra, M., "An Evaluation of Verification Procedures for CFD Applications," paper presented at the 24th Symposium on Naval Hydrodynamics, Fukuoka, Japan, July 8–13, 2002.

11. Beck, J. V., Blackwell, B. F., and St. Clair, C. R., *Inverse Heat Conduction*, Wiley, New York, 1985.

7

DATA ANALYSIS, REGRESSION, AND REPORTING OF RESULTS

Once data are acquired during an experiment, we must decide how the data can most effectively be presented. Much of the effort that has gone into planning, designing, and running the experiment will be wasted if the results are presented only in tabular form. The relationships between variables and the scatter in the data are usually very difficult to see from tabular data listings. Usually, graphic and mathematical representations of the data are used in addition to tabular presentation.

As a minimum, a report presenting experimental data should contain the following tabular information for each result: the value (single reading or average), the systematic standard uncertainty, the random standard uncertainty, and the expanded uncertainty with the level of confidence indicated. The data also should usually be presented in graphic form with the uncertainty bands around the results clearly indicated. We also often wish to obtain and report a mathematical expression to represent the data we have obtained. An equation is a much more compact representation of the data than tabular listings or even graphic representations, and it allows us to interpolate between the discrete data points.

In the following sections we consider how to obtain a mathematical expression to represent experimental results and what sort of uncertainty should be associated with such an expression. We then illustrate, using an example, how experimental results (and the uncertainties) might be presented using a regression equation. In the final section we discuss ideas associated with multiple linear regression.

7-1 OVERVIEW OF REGRESSION ANALYSIS AND ITS UNCERTAINTY[1]

When experimental information is represented by a regression, the regression model will have an associated uncertainty due to the uncertainty in the original experimental program. Consider the general case where a test is conducted on a hardware component. The data are plotted and a linear regression is used to determine the best fit of a curve through the data. (Note that the term *linear regression* means that the regression coefficients $a_0, a_1, \ldots, a_n$ are not functions of the X variable, not that the relationship between X and Y is linear.) Since the data are obtained experimentally, both X and Y will have experimental uncertainties, and these uncertainties will be made up of systematic and random uncertainties. A linear regression is then performed on the (X, Y) data and the general form of the regression is

$$Y(X) = a_0 + a_1 X + a_2 X^2 + \cdots + a_n X^n \qquad (7.1)$$

This polynomial equation is then used to represent the performance of the hardware. The original data and their uncertainty are usually forgotten in subsequent use. This model will not predict the *true* performance of the hardware because of the experimental uncertainties; thus the uncertainty associated with the regression is needed to provide the interval within which the *true* performance can be expected at a given level of confidence.

Obviously, an additional error is introduced if the wrong regression model is used, for example, if a third-order regression model is used and the true relationship is first order. This is the classical problem of incorrectly fitting the data, overfitting, or underfitting. *The error introduced by choice of an inappropriate regression model is not included in the methods discussed in this chapter.* It is assumed in the techniques presented that the correct model is being used and the only uncertainties in the regression model are those due to the uncertainties in the original experimental information.

The methodology presented in this chapter was evaluated and validated using Monte Carlo–type simulations [1, 2]. The following sections discuss uncertainty aspects of regression models and some guidance on reporting the uncertainty associated with regressions. As an example of the application of this methodology, the determination of the uncertainty in the calibration of a venturi flowmeter, and the uncertainty associated with its subsequent use for measurement of flow rate in a test are discussed.

7-1.1 Categories of Regression Uncertainty

Three primary categories can be defined based on how the regression information is being used. The first category is when the regression coefficients are the

---

[1] Section 7-1 is adapted from Refs. 1 and 2.

primary interest, the second category is when the regression model is used to provide a value for the Y variable, and the third category is when the (X_i, Y_i) data points are not measured quantities but are functions of other variables. This third category can include both of the other two categories, with the uncertainty associated with either the regression coefficients or the regression value of Y being of interest.

The key to the proper estimation of the uncertainty associated with a regression is a careful, comprehensive accounting of systematic uncertainties and correlated systematic errors. Correlated systematic errors will be present and must be accounted for properly when the (X_i, Y_i) and the (X_{i+1}, Y_{i+1}) data pairs have systematic errors from the same source. The examples presented below demonstrate some ways in which correlated systematic errors can be present.

7-1.2 Uncertainty in Coefficients

The general expression for a straight-line regression is

$$Y(X) = mX + c \tag{7.2}$$

where m is the slope of the line and c is the y intercept. In some experiments these coefficients are the desired information. An example is the stress–strain relationship for a linearly elastic material,

$$\sigma = E\epsilon \tag{7.3}$$

where the stress σ is linearly proportional to the strain ϵ by Young's modulus E. Young's modulus for a material is determined by measuring the elongation of the material for an applied load, calculating the normal stress and strain, and determining the slope of the line in the linearly elastic region. Since the stress and strain are determined experimentally, they will have experimental uncertainties, and thus the experimental value of Young's modulus will have an associated uncertainty.

7-1.3 Uncertainty in Y from Regression Model

Often, the uncertainty associated with a value determined using the regression model, Eq. (7.1) or (7.2), is desired. One would obtain a regression model for a given set of (X_i, Y_i) data, then use that regression model to obtain a Y value at a measured or specified X. The nature of the data and the use of the data determine how the uncertainty estimate is determined, such as:

1. Some or all (X_i, Y_i) data pairs from different experiments
2. All (X_i, Y_i) data pairs from the same experiment
3. New X from same apparatus

4. New X from different apparatus

5. New X with no uncertainty

It is instructive to discuss an example of each of these situations.

Suppose that heat transfer coefficient data sets from various facilities were combined as a single data set, with each facility contributing data over a slightly different range, and a regression model generated. The random and systematic standard uncertainties for each test apparatus are different. If the systematic standard uncertainties for the (X_i, Y_i) data and the (X_{i+1}, Y_{i+1}) data are obtained from the same apparatus and thus share the same error sources, their systematic errors will be correlated. However, if they are from different facilities and do not share any error sources, they will not be correlated. The uncertainty associated with the regression model must properly account for the correlation of the systematic errors.

The calibration of a thermocouple is an example where all the data could come from the same experiment. A calibration curve for the thermocouple would be generated by measuring an applied temperature and an output voltage, both of which contain uncertainties. The calibration curve could have the form

$$T = mE + c \tag{7.4}$$

where m and c are the regression coefficients determined from the (E_i, T_i) calibration data. When the thermocouple is then used in an experiment, a new voltage, E_{new}, is obtained. The new temperature is then found using the calibration curve:

$$T_{new} = mE_{new} + c \tag{7.5}$$

The uncertainty in T_{new} includes the uncertainty in the calibration curve as well as the uncertainty in the voltage measurement, E_{new}. If the same voltmeter is used in the experiment as was used in the calibration, the systematic error from the new voltage measurement will be correlated with the systematic error of each E_i used in finding the regression, and appropriate correlation terms are needed. If a different voltmeter is used to measure the new voltage, the systematic error of E_{new} will not be correlated with the systematic error of the E_i.

When the regression is used to represent a set of data and that regression is then used in an analysis, often the new X value is postulated and can be considered to have no uncertainty. An example would be pumping power P versus pump speed N for a centrifugal pump. The regression might have the form

$$P = a(N)^b \tag{7.6}$$

If an analyst uses this expression to obtain a value of power at a postulated value of N, the uncertainty in N could be considered to be zero.

7-1.4 (X_i, Y_i) Variables Are Functions

Another common situation is when the (X_i, Y_i) variables used in the regression are not the measured data but are each functions of several measured variables. The Young's modulus determination discussed previously is a typical example. Neither the stress nor the strain is measured directly. The stress is calculated from measurement of the applied force with a load cell and measurement of the cross-sectional area. The data reduction equation for stress is

$$\sigma = \frac{C V_{lc}}{V_i A} \tag{7.7}$$

where C is the calibration constant, V_{lc} is the load cell voltage, V_i is the excitation voltage, and A is the cross-sectional area. The strain is determined using a strain gage, and the data reduction equation is

$$\epsilon = \frac{2V_{br}}{G V_i} \tag{7.8}$$

where V_{br} is the bridge voltage, G is the gauge factor, and V_i is again the excitation voltage. The regression coefficient representing Young's modulus is thus a function of the variables C, V_{lc}, V_i, A, and G. In instances where error sources are shared between different variables, as would exist if all the voltages are measured with the same voltmeter, additional terms in the uncertainty propagation expression are necessary to account properly for the correlated systematic errors.

7-2 LEAST-SQUARES ESTIMATION

Linear regression analysis is based on minimizing the sum of the squares of the Y deviations between the line and the data points, commonly known as the *method of least squares*. Linear regression analysis can be divided into three broad categories: straight-line regressions, polynomial regressions, and multiple linear regressions. Multiple linear regressions are discussed briefly in Section 7-8. Straight-line regressions, also called first-order regressions or simple linear regressions, are a commonly used form. As discussed in Section 5-7, it is often recommended that if the data are not inherently linear, a transformation be used to try to obtain a linear relationship. Exponential functions and power law functions, for example, can be transformed to linear functions by appropriate logarithmic transformations. Reciprocal transformations are also very useful in linearizing a nonlinear function. If a suitable transformation cannot be found, a polynomial regression is then often used.

Consider a set of data points presented on an $X-Y$ plane. Here, X and Y may be variables in the experiment, some function of the variables ($Y = \ln y$, for instance) as discussed above, or nondimensional groups of variables (such as Nusselt number, Reynolds number, drag coefficient, etc.). Assume that all the uncertainty is concentrated in Y and that the uncertainty in the Y variable is the same for each value of X. Also assume that the "true" relationship between X and Y is a straight line.

We can now assume the general linear equation

$$Y_0 = mX + c \tag{7.9}$$

where Y_0 is taken as the "optimum" Y value to represent the data at a given X. We wish to find the values of the slope m and intercept c that minimize

$$\eta = \sum_{i=1}^{N} (Y_i - Y_0)^2 \tag{7.10}$$

where the Y_i are the N data values. Using Eq. (7.9), the expression to minimize becomes

$$\eta = \sum_{i=1}^{N} (Y_i - mX_i - c)^2 \tag{7.11}$$

For η to be minimum, we must find m and c such that

$$\frac{\partial \eta}{\partial m} = 0 \tag{7.12}$$

$$\frac{\partial \eta}{\partial c} = 0 \tag{7.13}$$

Taking the indicated derivatives and solving for m and c lead to

$$m = \frac{N \sum_{i=1}^{N} X_i Y_i - \sum_{i=1}^{N} X_i \sum_{i=1}^{N} Y_i}{N \sum_{i=1}^{N} (X_i^2) - \left(\sum_{i=1}^{N} X_i \right)^2} \tag{7.14}$$

$$c = \frac{\sum_{i=1}^{N} (X_i^2) \sum_{i=1}^{N} Y_i - \sum_{i=1}^{N} X_i \sum_{i=1}^{N} (X_i Y_i)}{N \sum_{i=1}^{N} (X_i^2) - \left(\sum_{i=1}^{N} X_i \right)^2} \tag{7.15}$$

The equation that results from this procedure Eq. (7.2),

$$Y = mX + c$$

is called a *regression equation*, and the procedure itself is sometimes referred to as "making a regression of Y on X." One point that should be emphasized concerns the intercept c. In many cases, we know what the intercept should be if the experiment were "perfect." Often this perfect intercept is $Y = 0$ at $X = 0$. This perfect value of the intercept should never be forced on the data. As Schenck

[3, p. 237] states, "The intercept of straight-line data is always inherent in the data and should be allowed to express itself."

One of the assumptions in the derivation of these expressions for the regression coefficients was that the uncertainty in the Y variable was constant over the range of the curve fit. This assumption is often violated in real engineering tests. Even though the data uncertainty varies, we will usually find that the least-squares regression will provide a mathematical expression that represents the data well. Since our purpose in using the regression is to obtain a mathematical expression that can be used to represent the data in a compact fashion (and perhaps to interpolate between discrete data points), the ultimate test is comparison of the regression curve with the data. If the data are well represented, the regression procedure can be considered successful.

Another restriction in the least-squares analysis presented above is that all the uncertainty is concentrated in Y. It is assumed that there is no uncertainty in the X values. This assumption is almost always violated since both X and Y are usually measured values with the possibility of both having random and systematic uncertainties.

The classical approach to regression uncertainty summarized below considers only the cases where there is no uncertainty in X. This approach also deals only with the random uncertainty component. The more realistic and general case is presented in Section 7-4, where the possibility of random and systematic uncertainties in X and Y is considered along with the correlated systematic errors that exist between the data points.

7-3 CLASSICAL LINEAR REGRESSION UNCERTAINTY: RANDOM UNCERTAINTY

Consider that a linear least-squares curve fit [Eq. (7.2)] has been made for a set of data. We can then form a statistic called the *standard error of regression* [4],

$$s_Y = \left[\frac{\sum_{i=1}^{N}(Y_i - mX_i - c)^2}{N - 2} \right]^{1/2} \tag{7.16}$$

which has the general form of a standard deviation. This quantity is an estimate of the standard deviation of the Y_i values for no uncertainty in the X_i values. The 2 subtracted from N in the denominator arises because two degrees of freedom are lost from the set of N data pairs (X_i, Y_i) when the curve-fit constants m and c are determined.

When we fit a least-squares regression through a set of data pairs, the resulting curve fit is essentially a relationship between the "mean" of Y and X. When we consider the expanded random uncertainty associated with the mean Y value obtained from the curve for a given value of X, we are looking for the interval that will contain the parent population mean of Y, μ_Y, with some level of confidence.

This interval for a 95% level of confidence is defined by [4]

$$Y - 2\left\{ s_Y^2 \left[\frac{1}{N} + \frac{(X - \overline{X})^2}{s_{XX}} \right] \right\}^{1/2} \leq \mu_Y \leq Y + 2\left\{ s_Y^2 \left[\frac{1}{N} + \frac{(X - \overline{X})^2}{s_{XX}} \right] \right\}^{1/2}$$

(7.17)

where

$$\overline{X} = \frac{1}{N} \sum_{i=1}^{N} X_i$$

(7.18)

and

$$s_{XX} = \sum_{i=1}^{N} X_i^2 - \frac{\left(\sum_{i=1}^{N} X_i \right)^2}{N}$$

(7.19)

and where Y is the value from the curve for a specific X. For no random uncertainty in X, Eq. (7.17) gives the expanded random uncertainty in the Y from the regression model. The factor 2 in Eq. (7.17) is the t value for $N - 2$ degrees of freedom and a 95% level of confidence where it is assumed that the large-sample approximation applies.

Even though the least-squares expression represents the mean Y, we sometimes use a curve fit to simplify the use of data, as in a calibration curve. In this case, the Y_i data values are considered more correct than the curve-fit values of Y. A value from the curve is used as an estimate of the value of Y_i at a given X. Since s_Y from Eq. (7.16) is the estimate of the standard deviation for the Y_i values, the appropriate random standard uncertainty s_{cal} to assign to a value from the calibration curve is

$$s_{cal} = s_Y$$

(7.20)

This random standard uncertainty interval around the curve-fit value for a given X should contain the "true" value of Y_i. This random standard uncertainty will become a fossilized systematic standard uncertainty. It represents only the curve-fit uncertainty and not any of the other uncertainties in the measurement, such as calibration standard, data acquisition, installation, and conceptual uncertainties.

Classical expressions are also available for the standard deviations for the slope and intercept from the curve fit [4]. The standard error (standard deviation) for the slope is

$$s_m = \left(\frac{s_Y^2}{s_{XX}} \right)^{1/2}$$

(7.21)

and that for the intercept is

$$s_c = \left[s_Y^2 \left(\frac{1}{N} + \frac{\overline{X}^2}{s_{XX}} \right) \right]^{1/2}$$

(7.22)

Of course, these expressions apply only if there is no random uncertainty in X. Using these standard deviations, the 95% confidence interval for the parent population mean values for the slope and intercept are

$$m - 2s_m \leq \mu_m \leq m + 2s_m \qquad (7.23)$$

and

$$c - 2s_c \leq \mu_c \leq c + 2s_c \qquad (7.24)$$

where large degrees of freedom are assumed.

These classical expressions for regression uncertainty presented in this section apply only when all the random uncertainty is concentrated in the Y values. However, even when this restriction applies, these expressions do not give the total uncertainty when there are systematic uncertainties in the data. In most cases the methodology given in the next section will be required to properly determine the uncertainties from a regression analysis.

7-4 COMPREHENSIVE APPROACH TO LINEAR REGRESSION UNCERTAINTY[2]

7-4.1 Uncertainty in Coefficients: First-Order Regression

As discussed in Section 7-2, for N (X_i, Y_i) data pairs, the slope m is determined from Eq. (7.14),

$$m = \frac{N \sum_{i=1}^{N} X_i Y_i - \sum_{i=1}^{N} X_i \sum_{i=1}^{N} Y_i}{N \sum_{i=1}^{N} \left(X_i^2\right) - \left(\sum_{i=1}^{N} X_i\right)^2}$$

and the intercept is determined from Eq. (7.15),

$$c = \frac{\sum_{i=1}^{N} \left(X_i^2\right) \sum_{i=1}^{N} Y_i - \sum_{i=1}^{N} X_i \sum_{i=1}^{N} \left(X_i Y_i\right)}{N \sum_{i=1}^{N} \left(X_i^2\right) - \left(\sum_{i=1}^{N} X_i\right)^2}$$

Considering Eqs. (7.14) and (7.15) to be data reduction equations of the form

$$m = m(X_1, X_2, \ldots, X_N, Y_1, Y_2, \ldots, Y_N) \qquad (7.25)$$

and

$$c = c(X_1, X_2, \ldots, X_N, Y_1, Y_2, \ldots, Y_N) \qquad (7.26)$$

---

[2]Section 7-4 is adapted from Refs. 1 and 2.

and applying the uncertainty analysis equations, the most general form of the expression for the 95% expanded uncertainty in the slope is

$$
U_m^2 = 2^2 \left[\sum_{i=1}^N \left(\frac{\partial m}{\partial Y_i} \right)^2 s_{Y_i}^2 + \sum_{i=1}^N \left(\frac{\partial m}{\partial X_i} \right)^2 s_{X_i}^2 + \sum_{i=1}^N \left(\frac{\partial m}{\partial Y_i} \right)^2 b_{Y_i}^2 \right.
$$
$$
+ 2 \sum_{i=1}^{N-1} \sum_{k=i+1}^N \left(\frac{\partial m}{\partial Y_i} \right) \left(\frac{\partial m}{\partial Y_k} \right) b_{Y_i Y_k} + \sum_{i=1}^N \left(\frac{\partial m}{\partial X_i} \right)^2 b_{X_i}^2
$$
$$
\left. + 2 \sum_{i=1}^{N-1} \sum_{k=i+1}^N \left(\frac{\partial m}{\partial X_i} \right) \left(\frac{\partial m}{\partial X_k} \right) b_{X_i X_k} + 2 \sum_{i=1}^N \sum_{k=1}^N \left(\frac{\partial m}{\partial X_i} \right) \left(\frac{\partial m}{\partial Y_k} \right) b_{X_i Y_k} \right]
$$

$$(7.27)$$

where s_{Y_i} is the random standard uncertainty for the Y_i variable, s_{X_i} is the random standard uncertainty for the X_i variable, b_{Y_i} is the systematic standard uncertainty for the Y_i variable, b_{X_i} is the systematic standard uncertainty for the X_i variable, $b_{Y_i Y_k}$ is the covariance estimator for the correlated systematic errors in the Y_i and Y_k variables, $b_{X_i X_k}$ is the covariance estimator for correlated systematic errors in the X_i and X_k variables, and $b_{X_i Y_k}$ is the covariance estimator for the correlated systematic errors between X_i and Y_k.

A similar expression for the 95% expanded uncertainty in the intercept is

$$
U_c^2 = 2^2 \left[\sum_{i=1}^N \left(\frac{\partial c}{\partial Y_i} \right)^2 s_{Y_i}^2 + \sum_{i=1}^N \left(\frac{\partial c}{\partial X_i} \right)^2 s_{X_i}^2 + \sum_{i=1}^N \left(\frac{\partial c}{\partial Y_i} \right)^2 b_{Y_i}^2 \right.
$$
$$
+ 2 \sum_{i=1}^{N-1} \sum_{k=i+1}^N \left(\frac{\partial c}{\partial Y_i} \right) \left(\frac{\partial c}{\partial Y_k} \right) b_{Y_i Y_k} + \sum_{i=1}^N \left(\frac{\partial c}{\partial X_i} \right)^2 b_{X_i}^2
$$
$$
\left. + 2 \sum_{i=1}^{N-1} \sum_{k=i+1}^N \left(\frac{\partial c}{\partial X_i} \right) \left(\frac{\partial c}{\partial X_k} \right) b_{X_i X_k} + 2 \sum_{i=1}^N \sum_{k=1}^N \left(\frac{\partial c}{\partial X_i} \right) \left(\frac{\partial c}{\partial Y_k} \right) b_{X_i Y_k} \right]
$$

$$(7.28)$$

These equations show the most general form of the expressions for the uncertainty in the slope and intercept, allowing for correlation of systematic errors among the different values of X, among the different values of Y, and also among the values of X and Y. If none of the systematic error sources are common between the X and Y variables, the last term of Eqs. (7.27) and (7.28), the $X-Y$ covariance estimator, is zero. Also, if the random standard uncertainties s_{Y_i} are constant over the range of the curve fit, the random standard uncertainties determined for m and c from the first summation on the RHS of Eqs. (7.27) and (7.28) are the same as those determined using the classical method, Eqs. (7.21) and (7.22). (Remember that the partial derivatives can be determined numerically. It is not necessary to find analytical expressions for them.)

7-4.2 Uncertainty in Y from Regression Model: First-Order Regression

The expression for the 95% expanded uncertainty in the Y determined from the regression model at a measured or specified value of X is found by substituting Eqs. (7.14) and (7.15) into

$$Y(X_{\text{new}}) = mX_{\text{new}} + c \qquad (7.29)$$

and applying the uncertainty propagation equation to obtain the uncertainty in Y as

$$
\begin{aligned}
U_Y^2 = 2^2 &\left[\sum_{i=1}^{N} \left(\frac{\partial Y}{\partial Y_i} \right)^2 s_{Y_i}^2 + \sum_{i=1}^{N} \left(\frac{\partial Y}{\partial X_i} \right)^2 s_{X_i}^2 + \sum_{i=1}^{N} \left(\frac{\partial Y}{\partial Y_i} \right)^2 b_{Y_i}^2 \right. \\
&+ 2 \sum_{i=1}^{N-1} \sum_{k=i+1}^{N} \left(\frac{\partial Y}{\partial Y_i} \right) \left(\frac{\partial Y}{\partial Y_k} \right) b_{Y_i Y_k} + \sum_{i=1}^{N} \left(\frac{\partial Y}{\partial X_i} \right)^2 b_{X_i}^2 \\
&+ 2 \sum_{i=1}^{N-1} \sum_{k=i+1}^{N} \left(\frac{\partial Y}{\partial X_i} \right) \left(\frac{\partial Y}{\partial X_k} \right) b_{X_i X_k} + 2 \sum_{i=1}^{N} \sum_{k=1}^{N} \left(\frac{\partial Y}{\partial X_i} \right) \left(\frac{\partial Y}{\partial Y_k} \right) b_{X_i Y_k} \\
&+ \left(\frac{\partial Y}{\partial X_{\text{new}}} \right)^2 b_{X_{\text{new}}}^2 + \left(\frac{\partial Y}{\partial X_{\text{new}}} \right)^2 s_{X_{\text{new}}}^2 \\
&\left. + 2 \sum_{i=1}^{N} \left(\frac{\partial Y}{\partial X_{\text{new}}} \right) \left(\frac{\partial Y}{\partial X_i} \right) b_{X_{\text{new}} X_i} + 2 \sum_{i=1}^{N} \left(\frac{\partial Y}{\partial X_{\text{new}}} \right) \left(\frac{\partial Y}{\partial Y_i} \right) b_{X_{\text{new}} Y_i} \right]
\end{aligned}
$$
$$(7.30)$$

The first seven terms on the RHS of Eq. (7.30) account for uncertainties from the (X_i, Y_i) data pairs. The eighth and ninth terms account for the systematic and random uncertainties for the new X. The tenth term is included if the new X variable is measured with the same apparatus as that used to measure the original X-variables. The last term is included if error sources are common between the new X and the original Y_i variables. As with the regression coefficients, if the s_{Y_i} uncertainties are constant over the range of the curve fit, the random standard uncertainty for Y in Eq. (7.30) is the same as that obtained from the term in braces in Eq. (7.17).

If an analysis is being conducted using a Y value determined using Eq. (7.29) and it is postulated that the X value being used in the regression model, X_{new}, has no uncertainty, terms 8 through 11 on the RHS of Eq. (7.30) are omitted. An example of a first-order regression and the determination of its uncertainty are presented in Section 7-7 for the calibration of a pressure transducer.

A special case for the systematic standard uncertainty part of Eq. (7.30) that applies in many first-order regression models, especially calibration curves, is described in Tables 7.1 and 7.2 [5]. If all the Y_i data points have the same systematic standard uncertainty b_y with all the systematic errors for the Y_i's

Table 7.1 Systematic Standard Uncertainty *Components* for Y Determined from First-Order Regression Equation (7.30) When Each Y_i Has the Same Systematic Standard Uncertainty b_y and Each X_i Has the Same Systematic Standard Uncertainty b_x [5]

Systematic standard uncertainty in Y_i data	$b_{Y_1} = b_y$
Systematic standard uncertainty in X_i data with no systematic standard uncertainty in X_{new}	$b_{Y_2} = mb_x$
Systematic standard uncertainty in X_i data with correlated systematic errors in X_{new}	0
Systematic standard uncertainty in X_i data with uncorrelated systematic errors in X_{new}	$b_{Y_3} = [(mb_x)^2 + (mb_{X_{new}})^2]^{1/2}$

Table 7.2 Systematic Standard Uncertainty for Y Determined from First-Order Regression Equation (7.30) When Each Y_i Has the Same Systematic Standard Uncertainty b_y and Each X_i Has the Same Systematic Standard Uncertainty b_x [5]

Systematic standard uncertainty in Y_i data only	$b_Y = b_{Y_1}$
Systematic standard uncertainty in Y_i data and in X_i data	
Systematic standard uncertainty in X_i data with no systematic standard uncertainty in X_{new}	$b_Y = \sqrt{(b_{Y_1})^2 + (b_{Y_2})^2}$
Systematic standard uncertainty in X_i data with correlated systematic errors in X_{new}	$b_Y = b_{Y_1}$
Systematic standard uncertainty in X_i data with uncorrelated systematic errors in X_{new}	$b_Y = \sqrt{(b_{Y_1})^2 + (b_{Y_3})^2}$

correlated, and if all the X_i data points have the same systematic standard uncertainty b_x with all the systematic errors correlated, then the systematic standard uncertainty terms in Eq. (7.30) reduce to the components given in Table 7.1.

The first expression in Table 7.1 is the systematic standard uncertainty in the Y determined from the regression model because of the systematic uncertainty in the Y_i data. The second expression accounts for the systematic standard uncertainty in Y resulting from the systematic uncertainty in the X_i data when there is no systematic uncertainty in the X_{new} data. When there are systematic errors in X_{new} and these errors are correlated with those in the X_i data, the result is a cancellation of the effects of these systematic errors on the Y result. The last expression in Table 7.1 applies when the X_i data and X_{new} have independent systematic errors.

The resulting systematic standard uncertainty for Y for this special case is shown in Table 7.2. Depending on which of the components from Table 7.1 apply, these components are root sum squared to yield the systematic standard uncertainty for Y. This value is then combined with the appropriate random standard uncertainty components in Eq. (7.30) and expanded to the desired level of confidence.

7-4.3 Higher Order Regressions

The general expression for an nth-order polynomial regression model is

$$Y(X_{\text{new}}) = a_0 + a_1 X_{\text{new}} + a_2 X_{\text{new}}^2 + \cdots + a_n X_{\text{new}}^n \qquad (7.31)$$

where the regression coefficients a_i are determined with a least-squares fit similar to that used for the first-order case [6]. The expression for the uncertainty in a value found using the polynomial regression model is the same as Eq. (7.30), with Y defined by Eq. (7.31). The complexity of the polynomial regression determination usually makes analytical determination of the partial derivatives prohibitive.

7-5 REPORTING REGRESSION UNCERTAINTIES[3]

After the uncertainty associated with a regression has been calculated, it should be documented clearly and concisely so that it can be easily used. Figure 7.1 shows a set of (X, Y) data, the first-order regression model for that data set, and the associated uncertainty interval determined using Eq. (7.30). It would usually be very useful to have not only the regression model but also an equation giving

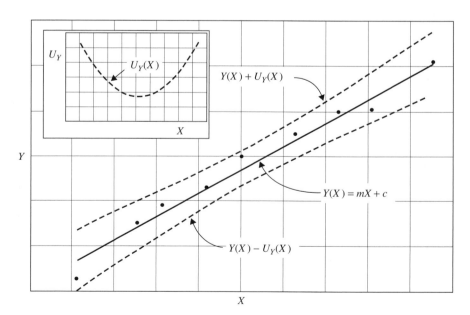

Figure 7.1 Expressing regression uncertainty intervals. (From Refs. 1 and 2.)

[3]Section 7-5 is adapted from Refs. 1 and 2.

$U_Y(X)$. While Eq. (7.30) gives such information, it requires having the entire (X_i, Y_i) data set as input each time a calculation of U_Y is made for a new X value because the partial derivatives in Eq. (7.30) are functions of X_{new}. Since this is cumbersome and inconvenient, the authors propose [2] that a set of (X, U) points be calculated using Eq. (7.30) and a regression be performed using these points to produce an expression for $U_{Y\text{-regress}}(X)$, which is used as explained below to calculate $U_Y(X)$ at a given X value.

The procedure that should be used is as follows. If the X_{new} to be used in Eq. (7.1),

$$Y(X) = a_0 + a_1 X + a_2 X^2 + \cdots + a_n X^n$$

has no uncertainty, then Eq. (7.30)—with terms 8 through 11 on the RHS set equal to zero—is used over the desired X range to produce the (X, U) data set to be curve fit to produce $U_{Y\text{-regress}}(X)$.

If the X_{new} to be used in Eq. (7.1) does have uncertainty and is from the same apparatus as produced the (X_i, Y_i) data set, all of the terms in Eq. (7.30) must be considered. However, the $s_{X_{new}}^2$ term will often be larger during the test than during the more well-controlled calibration. This can be taken into account by assigning the same systematic uncertainty components that were in the X_i data to the X_{new} values used in Eq. (7.30) to produce the (X, U) points to curve fit but with the $s_{X_{new}}^2$ term in Eq. (7.30) set equal to zero. The resulting (X, U) points are then curve fit to yield $U_{Y\text{-regress}}(X)$. When an X_{new} is used in Eq. (7.1), the uncertainty U_Y in the Y determined from the regression is then calculated as

$$(U_Y)^2 = U_{Y\text{-regress}}^2 + 2^2 \left(\frac{\partial Y}{\partial X_{new}} \right)^2 s_{X_{new}}^2 \tag{7.32}$$

If the X_{new} to be used in Eq. (7.1) is from a different apparatus than that which produced the (X_i, Y_i) data set, terms 10 and 11 on the RHS of Eq. (7.30) will be zero. This situation would be encountered if a thermocouple calibration curve was determined but then another voltmeter was used in testing rather than that which produced the original (T_i, E_i) data set. Equation (7.30)—with terms 8 through 11 on the RHS set equal to zero—is used over the desired X range to calculate a set of (X, U) points that are then curve fit to produce a $U_{Y\text{-regress}}(X)$ expression. When an X_{new} is used in Eq. (7.1), the uncertainty U_Y in the determined Y is then calculated using

$$(U_Y)^2 = U_{Y\text{-regress}}^2 + 2^2 \left(\frac{\partial Y}{\partial X_{new}} \right)^2 (b_{X_{new}}^2 + s_{X_{new}}^2) \tag{7.33}$$

The contribution of $U_{Y\text{-regress}}$ to U_Y in Eqs. (7.32) and (7.33) is purely a systematic uncertainty, since all random uncertainties are fossilized once the regression is performed. It should be noted that the lowest order curve fit that provides an acceptable fit for the regression uncertainty $U_{Y\text{-regress}}$ should be used to represent that uncertainty.

7-6 REGRESSIONS IN WHICH X AND Y ARE FUNCTIONAL RELATIONS[4]

In many, if not most, instances the test results will be expressed in terms of functional relations, often dimensionless. In these cases the measured variables will not be the X values and Y values used in the regression. Examples of common functional relations are Reynolds number, flow coefficient, turbine efficiency, and specific fuel consumption. For example, in the experimental determination of the pressure distribution on the surface of a cylinder in crossflow, the results are usually presented as pressure coefficient versus angular position θ. The pressure coefficient is

$$C_P = \frac{P - P_\infty}{q_\infty} \tag{7.34}$$

where P is the local pressure, P_∞ is the free-stream pressure, and q_∞ is the dynamic pressure. The Y_i variables used in the regression model are the determined pressure coefficient values and are functions of the three measured variables

$$Y_i = C_{p,i} = f(P_i, P_{\infty,i}, q_{\infty,i}) \tag{7.35}$$

and the X_i variables are the angular positions θ_i. If the pressures are measured with separate transducers that were calibrated against the same standard, the systematic errors from the calibration will be correlated and must be properly taken into account. Equations (7.27), (7.28), and (7.30) apply in such cases, with Y (in this case C_p) taken as the dependent function for the regression and the X_i values and Y_i values replaced by the full set of measured variables (in this case the P_i, $P_{\infty,i}$, $q_{\infty,i}$, and θ_i values).

In general, for an experiment with N sets of J measured variables, the regression model can be viewed as being a function of the J variables, such as

$$Y(X_{\text{new}}, X_i, Y_i) = f(\text{VAR}1_i, \text{VAR}2_i, \ldots, \text{VAR}J_i) \tag{7.36}$$

where the function X_i is made up of some of the variables and the function Y_i is made up of some of the variables, such as

$$X_i = f(\text{VAR}1_i, \text{VAR}2_i, \ldots, \text{VAR}K_i, \ldots) \tag{7.37}$$

---

[4]Section 7-6 is adapted from Refs. 1 and 2.

and

$$Y_i = f(\ldots, \text{VAR}K_i, (\text{VAR}K + 1)_i, \ldots, \text{VAR}J_i) \tag{7.38}$$

and X_{new} is the new value of the independent variable in the regression. The general expression for the uncertainty in the value determined from the regression model becomes

$$
\begin{aligned}
U_Y^2 = 2^2 \Bigg[& \sum_{i=1}^{N} \sum_{j=1}^{J} \left(\frac{\partial Y}{\partial \text{VAR} j_i} \right)^2 s_{\text{VAR} j_i}^2 \\
& + \sum_{i=1}^{N} \sum_{j=1}^{J} \left(\frac{\partial Y}{\partial \text{VAR} j_i} \right)^2 b_{\text{VAR} j_i}^2 \\
& + 2 \sum_{i=1}^{N-1} \sum_{k=i+1}^{N} \sum_{j=1}^{J} \left(\frac{\partial Y}{\partial \text{VAR} j_i} \right) \left(\frac{\partial Y}{\partial \text{VAR} j_k} \right) b_{\text{VAR} j_i \, \text{VAR} j_k} \\
& + 2 \sum_{i=1}^{N} \sum_{k=1}^{N} \sum_{j=1}^{J-1} \sum_{l=j+1}^{J} \left(\frac{\partial Y}{\partial \text{VAR} j_i} \right) \left(\frac{\partial Y}{\partial \text{VAR} l_k} \right) b_{\text{VAR} j_i \, \text{VAR} l_k} \\
& + \sum_{j=1}^{J} \left(\frac{\partial Y}{\partial \text{VAR} j_{\text{new}}} \right)^2 s_{\text{VAR} j_{\text{new}}}^2 + \sum_{j=1}^{J} \left(\frac{\partial Y}{\partial \text{VAR} j_{\text{new}}} \right)^2 b_{\text{VAR} j_{\text{new}}}^2 \\
& + 2 \sum_{i=1}^{N} \sum_{j=1}^{J} \left(\frac{\partial Y}{\partial \text{VAR} j_{\text{new}}} \right) \left(\frac{\partial Y}{\partial \text{VAR} j_i} \right) b_{\text{VAR} j_{\text{new}}} b_{\text{VAR} j_i} \\
& + 2 \sum_{i=1}^{N} \sum_{j=1}^{J-1} \sum_{l=j+1}^{J} \left(\frac{\partial Y}{\partial \text{VAR} j_{\text{new}}} \right) \left(\frac{\partial Y}{\partial \text{VAR} l_i} \right) b_{\text{VAR} j_{\text{new}}} b_{\text{VAR} l_i} \Bigg] \tag{7.39}
\end{aligned}
$$

where the third RHS term accounts for correlated systematic error sources within each variable and the fourth RHS term accounts for systematic error sources common between variables. Similar expressions for the uncertainty in the slope and intercept are readily obtained. Since the X_i and Y_i variables used to determine the regression model are now functional relations, the determination of the partial derivatives becomes more complex and should probably be performed numerically.

Linearizing transformations are often used to transform nonlinear functions to linear functions. For example, an exponential function is a linear function in semilogarithmic space. The uncertainty associated with the regression model can be estimated using the techniques discussed above, where the functional relationships, as expressed in Eqs. (7.37) and (7.38), are logarithmic functions. Again, the complexity introduced by the transformation usually makes analytical determination of the partial derivatives prohibitive.

7-7 EXAMPLES OF DETERMINING REGRESSIONS AND THEIR UNCERTAINTIES[5]

In this section we discuss the application of the regression uncertainty methodology to the calibration of a pressure transducer, which was then used in the calibration of a venturi flowmeter, which was subsequently used to determine water flow rate in a test [1, 2]. (It is shown that the frequently used way of accounting for the contribution of discharge coefficient uncertainty to the overall flow-rate uncertainty does not account correctly for all uncertainty sources, and the appropriate approach is presented and discussed.)

Venturi flowmeters are often used to measure fluid flow because of their simplicity, durability, and lack of moving parts. They are part of a class of flowmeters known as differential pressure producers since a pressure drop across a region in the meter is produced, and this pressure differential is used to determine the flow rate through the meter. The one-dimensional model for the volumetric flow rate through a venturi flowmeter is [7]

$$Q_{Th} = \frac{A_2}{\sqrt{1 - (A_2/A_1)^2}} \sqrt{2 \frac{P_1 - P_2}{\rho}} \qquad (7.40)$$

If the pressure drop measured between taps 1 and 2 is used to calculate the "head" of the flowing fluid,

$$h = \frac{P_1 - P_2}{\rho g} \qquad (7.41)$$

and the ratio of the venturi throat diameter d to the venturi inlet diameter D is denoted as

$$\beta = \frac{d}{D} \qquad (7.42)$$

then the one-dimensional flow rate can be expressed as

$$Q_{Th} = K d^2 \sqrt{\frac{h}{1 - \beta^4}} \qquad (7.43)$$

where K is a constant.

The actual flow rate will never equal the theoretical flow rate, so the venturi was calibrated against a reference meter. The discharge coefficient C_d is defined as the flow rate through the standard divided by the theoretical flow rate,

$$C_d = \frac{Q_{std}}{Q_{Th}} = \frac{Q_{std}}{K d^2 \sqrt{h/(1 - \beta^4)}} \qquad (7.44)$$

[5]Section 7-7 is adapted from Refs. 1 and 2.

Often, a venturi flowmeter is calibrated over a range of flow rates and the discharge coefficient is written as a function of Reynolds number defined at the venturi throat,

$$\text{Re} = \frac{Vd}{\nu} = \frac{4Q}{\pi d \nu} \qquad (7.45)$$

In this example, a first-order regression model

$$C_d(\text{Re}) = a_0 + a_1 \text{Re} \qquad (7.46)$$

is determined from the $N(\text{Re}(i), C_d(i))$ data points produced in the calibration. When the venturi flowmeter is used in a test, the flow rate is determined from

$$Q = C_d(\text{Re}_{new}) K d^2 \sqrt{\frac{h_{new}}{1 - (d/D)^4}} \qquad (7.47)$$

or, substituting Eqs. (7.45) and (7.46) into (7.47),

$$Q = \frac{a_0 K d^2 \sqrt{\dfrac{h_{new}}{1 - (d/D)^4}}}{1 - \dfrac{4 a_1 K d}{\pi \nu_{new}} \sqrt{\dfrac{h_{new}}{1 - (d/D)^4}}} \qquad (7.48)$$

Since the variables d, D, and possibly ν_{new} and h_{new} in Eqs. (7.47) and (7.48) share identical systematic error sources with the same variables in the $(\text{Re}(i), C_d(i))$ data points used in the regression that produced the a_0 and a_1 coefficients, the estimate of the uncertainty in Q should include correlated systematic error terms to take this into account. This is often not done, as we discuss further in a following section.

Note that the kinematic viscosity ν_{new} appearing in Re_{new} will be different than the $\nu(i)$ used in determining $\text{Re}(i)$ if the venturi is calibrated with a different fluid from that being used during the test. In such cases, since the systematic uncertainties do not arise from the same source, no correlation terms are needed.

7-7.1 Experimental Apparatus

A calibration apparatus (shown in Figure 7.2) was assembled to calibrate the 2-in.-diameter venturi flowmeter. A variable-speed pump was used to pump water through the flowmeters, the flow rate was measured with the reference flowmeter, and the differential pressure produced by the venturi flowmeter was recorded. The flowmeter used as the calibration standard was a $1\frac{1}{2}$-in.-diameter turbine-type flowmeter. This flowmeter was used because it was available and was installed

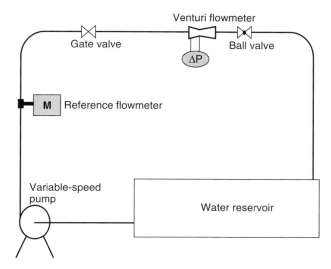

Figure 7.2   Schematic of experimental apparatus for venturi calibration. (From Ref. 1.)

in the water tunnel facility being used. The differential pressure was measured with a differential pressure transducer that required a 12-V dc excitation voltage and produced a millivolt output proportional to the pressure difference between the two ports on the transducer.

7-7.2 Pressure Transducer Calibration and Uncertainty

The transducer was calibrated by applying a water column to each port of the transducer and recording the output voltage as the height of the water column on one side was increased. The calibration data points for head and output voltage $[h_c(i), V_c(i)]$ are given in Table 7.3, and the first-order regression representing them,

$$h = 2.2624V_0 - 0.42334 \tag{7.49}$$

(with h in inches of H_2O and V_0 in millivolts), is shown in Figure 7.3. When the calibrated transducer is used in a test, a value of the output voltage V_0 is read and Eq. (7.49) is used to obtain the value of head. The uncertainty associated with the h value determined from this transducer calibration curve was assessed using Eq. (7.30) (where in this case $Y_i = h_c$ and $X_i = V_c$).

No error sources were shared between the measured differential pressure and the measured transducer output voltage, so all b_{XY} terms were zero. Since a random uncertainty estimate for X_{new} based on the transducer calibration data might not be representative of the random uncertainty in X_{new} encountered when the transducer is used in a test to determine the flow rate through the venturi,

Table 7.3 Pressure Transducer Calibration Data Set

Data Point (i)	$h_c(i)$(in. H_2O)[a]	$V_c(i)$ (mV)
1	0	0.3
2	1.875	1.0
3	4	2.1
4	8.25	3.9
5	13	6.0
6	17.25	7.8
7	22	9.9
8	23.75	10.7
9	26.125	11.7
10	26.5	11.9
11	31.375	14.0
12	33.375	15.0
13	35.875	16.0
14	40.375	18.0
15	41.75	18.6
16	44.875	20.1
17	49.25	22.0
18	49.875	22.2
19	53.875	24.0
20	55	24.5
21	58.375	26.0
22	60.75	27.1
23	61.875	27.5
24	65.625	29.2
25	68.25	30.4
26	68.125	30.4
27	72.625	32.1
28	77.125	34.3
29	−0.125	−0.3
30	8.125	3.8
31	14.25	6.6
32	22.625	10.2
33	31	13.9
34	41.375	18.5
35	46.625	20.8
36	52.5	23.4
37	58.75	26.1

[a]Resolution was $\frac{1}{8}$ in. H_2O.

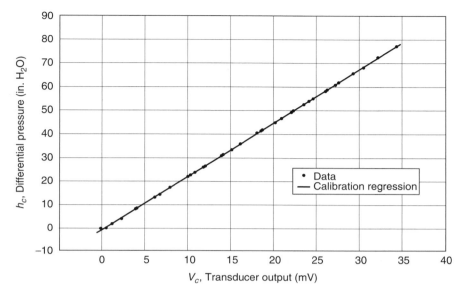

Figure 7.3 Differential pressure transducer calibration curve. (From Ref. 1.)

the $s_{X_{new}}$ term in Eq. (7.30) was omitted from the regression uncertainty and will be accounted for in the flow-rate uncertainty. [This corresponds to the case in which Eq. (7.32) is used to determine U_Y.] This is a common situation since the calibration process is usually better controlled and provides more stable data than when the transducer is used in the actual experiment.

The uncertainty contribution for the value of h determined from the differential pressure transducer calibration regression then becomes

$$
\begin{aligned}
U^2_{h\text{-regress}} = 2^2 & \left[\sum_{i=1}^{N} \left(\frac{\partial h}{\partial h_c(i)} \right)^2 s^2_{h_c(i)} + \sum_{i=1}^{N} \left(\frac{\partial h}{\partial V_c(i)} \right)^2 s^2_{V_c(i)} \right. \\
& + \sum_{i=1}^{N} \left(\frac{\partial h}{\partial h_c(i)} \right)^2 b^2_{h_c(i)} + 2 \sum_{i=1}^{N-1} \sum_{k=i+1}^{N} \left(\frac{\partial h}{\partial h_c(i)} \right) \\
& \times \left(\frac{\partial h}{\partial h_c(k)} \right) b_{h_c(i)h_c(k)} + \sum_{i=1}^{N} \left(\frac{\partial h}{\partial V_c(i)} \right)^2 b^2_{V_c(i)} \\
& + 2 \sum_{i=1}^{N-1} \sum_{k=i+1}^{N} \left(\frac{\partial h}{\partial V_c(i)} \right) \left(\frac{\partial h}{\partial V_c(k)} \right) b_{V_c(i)V_c(k)} \\
& \left. + \left(\frac{\partial h}{\partial V_0} \right)^2 b^2_{V_0} + 2 \sum_{i=1}^{N} \left(\frac{\partial h}{\partial V_0} \right) \left(\frac{\partial h}{\partial V_c(i)} \right) b_{V_0 V_c(i)} \right] \quad (7.50)
\end{aligned}
$$

A systematic standard uncertainty of $\frac{1}{32}$ in. H_2O and a random standard uncertainty of $\frac{1}{16}$ in. H_2O were used as the uncertainty estimates for the applied differential pressure. These estimates were based on the accuracy and readability of the scale used to measure the heights of the water columns. A systematic standard uncertainty of 0.25 mV was estimated for the voltmeter based on the manufacturer's specifications. The uncertainty associated with the differential pressure transducer calibration curve was obtained from Eq. (7.50) and is shown as a function of the output voltage in Figure 7.4. The second-order polynomial

$$U_{h\text{-regress}} = 5.28 \times 10^{-5} V_0^2 - 0.00177 V_0 + 0.0833 \qquad (7.51)$$

was fit to the uncertainty results and used to express the differential pressure uncertainty as a function of the transducer output voltage. As discussed previously, this expression does not include the contribution of the random uncertainty in the V_0 measured during the testing, so an equation similar to Eq. (7.32) must be used to estimate the overall U_h.

7-7.3 Venturi Discharge Coefficient and Its Uncertainty

As discussed earlier, approaches used often have not taken all of the correlated error contributions into account in estimating uncertainties in discharge

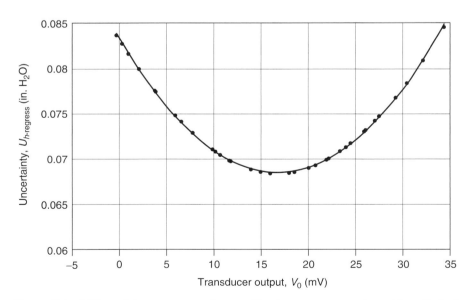

Figure 7.4   Differential pressure transducer calibration curve uncertainty. (From Refs. 1 and 2.)

coefficients or in estimating the flow-rate uncertainties. In this section, we present the approach for determining the uncertainty in the discharge coefficient, and we discuss the deficiencies in trying to use this information to find the uncertainty of the flow rate calculated from Eq. (7.47). In the next section, an approach [1, 2] that accounts properly for all correlated error contributions is demonstrated for the flow-rate determination.

In the calibration experiment discussed in this example, the flow rate Q_{std} measured by the calibration standard flowmeter and the differential pressure produced by the venturi flowmeter were obtained at selected settings of the variable-speed pump. Using Eqs. (7.44) and (7.45), obtaining the dimensions of the venturi from the manufacturer's drawings ($D = 2.125$ in., $d = 1.013$ in.) and a reference room temperature value for kinematic viscosity of water (1.08×10^{-5} ft^2/s), the Reynolds number and discharge coefficient for each data point were calculated.

The data set, given in Table 7.4, that was used to determine the regression model was obtained by averaging the Q_{std} and h data taken at each pump speed setting. The (Re(i), $C_d(i)$) data set was fit with a first-order regression model that yielded

$$C_d = 0.991 - 1.21 \times 10^{-7}\,\mathrm{Re} \tag{7.52}$$

Recalling Eqs. (7.44) and (7.45), the C_d predicted by this expression is actually a function of the variables $Q_{std}(i)$, $h(i)$, $v(i)$, d, and D and can be written functionally as

$$C_d = f\{Q_{std}(i), h(i), v(i), d, D\} \tag{7.53}$$

Note that $h(i)$ is a function of $V_0(i)$ through Eq. (7.49) and also a function of the head and voltage data used to obtain Eq. (7.49).

Table 7.4   Venturi Calibration Data Set

$Q_{std}(i)$ (gpm)	$V_0(i)$ (mV)	Re(i)	$C_d(i)$
20.35	5.45	63,553	0.987
23.55	7.30	73,546	0.982
26.50	9.30	82,759	0.977
33.20	14.47	103,683	0.977
39.20	20.23	122,420	0.974
45.17	26.93	141,064	0.972
50.68	33.63	158,272	0.975
56.15	41.18	175,354	0.976
61.70	49.63	192,688	0.976
67.08	58.45	209,488	0.978

The expression for the uncertainty associated with a predicted value from the discharge coefficient regression model is obtained by applying Eq. (7.39) to Eq. (7.53), yielding

$$
\begin{aligned}
U^2_{C_d\text{-regress}} = 2^2 \Bigg[& \sum_{i=1}^{N}\left(\frac{\partial C_d}{\partial Q_{\text{std}}(i)}\right)^2 s^2_{Q_{\text{std}}(i)} + \sum_{i=1}^{N}\left(\frac{\partial C_d}{\partial Q_{\text{std}}(i)}\right)^2 b^2_{Q_{\text{std}}(i)} \\
& + 2\sum_{i=1}^{N-1}\sum_{k=i+1}^{N}\left(\frac{\partial C_d}{\partial Q_{\text{std}}(i)}\right)\left(\frac{\partial C_d}{\partial Q_{\text{std}}(k)}\right) b_{Q_{\text{std}}(i)Q_{\text{std}}(k)} \\
& + \sum_{i=1}^{N}\left(\frac{\partial C_d}{\partial h(i)}\right)^2 [b^2_{h(i)}+s^2_{h(i)}] + 2\sum_{i=1}^{N-1}\sum_{k=i+1}^{N}\left(\frac{\partial C_d}{\partial h(i)}\right) \\
& \times \left(\frac{\partial C_d}{\partial h(k)}\right) b_{h(i)h(k)} + \sum_{i=1}^{N}\left(\frac{\partial C_d}{\partial v(i)}\right)^2 b^2_{v(i)} \\
& + 2\sum_{i=1}^{N-1}\sum_{k=i+1}^{N}\left(\frac{\partial C_d}{\partial v(i)}\right)\left(\frac{\partial C_d}{\partial v(k)}\right) b_{v(i)v(k)} + \left(\frac{\partial C_d}{\partial d}\right)^2 b^2_d \\
& + \left(\frac{\partial C_d}{\partial D}\right)^2 b^2_D \Bigg]
\end{aligned}
\tag{7.54}
$$

The $b_{h(i)}$ in Eq. (7.54) is the standard uncertainty equivalent of Eq. (7.50) since all uncertainty contributions to Eq. (7.50) are fossilized when it is used. The $s_{h(i)}$ in Eq. (7.54) is found by applying the uncertainty propagation equation to Eq. (7.49). Note that in Eq. (7.54) we have included no contributions from the uncertainties in Re_{new} when Eq. (7.52) is used.

The evaluation of correlation terms such as those for $h(i)h(k)$ and $Q_{\text{std}}(i)Q_{\text{std}}(k)$ requires further discussion. Since h is determined from Eq. (7.49) using a measured voltage V_0, the identical error source that causes the systematic errors in two sequential h determinations to be correlated occurs in the V_0 measurements. A similar situation occurs if Q_{std} is determined using a calibration equation and a measured frequency. Generalize these cases by considering a variable T that is determined using a calibration equation $T = f(E)$ and a measured value of the variable E. The covariance approximator $b_{T(i)T(j)}$ is given by

$$
b_{T(i)T(j)} = \left(\frac{\partial T}{\partial E}\right)_i \left(\frac{\partial T}{\partial E}\right)_j b_{E(i)E(j)}
\tag{7.55}
$$

where $b_{E(i)E(j)}$ is determined as usual from the sum-of-products method described in Chapter 5. Specifically, for $b_{h(i)h(k)}$ in Eq. (7.54) the form of this

Table 7.5   Uncertainties (95%) for Venturi Calibration

Source	Systematic	Random
Turbine meter	1% of reading	0.5 gpm
Differential pressure transducer	Eq. (7.51)	$2(\partial h/\partial V_0) \times s_{V_0}$
Venturi throat diameter	0.005 in.	0.0
Venturi inlet diameter	0.005 in.	0.0
Kinematic viscosity	0.5% of reading	0.0

term is

$$b_{h(i)h(k)} = \left(\frac{\partial h}{\partial V_0}\right)_{V_0(i)} \left(\frac{\partial h}{\partial V_0}\right)_{V_0(k)} b_{V_0(i)V_0(k)} \qquad (7.56)$$

The uncertainties estimated for the venturi discharge coefficient variables are shown in Table 7.5. The calibration flowmeter had an uncertainty quoted by the manufacturer of 1.0% of reading, and one-half of this value was used as the systematic standard uncertainty in Eq. (7.54). The calibration flowmeter random uncertainty was estimated based on variation observed during the testing. When the differential pressure transducer calibration curve, Eq. (7.49), is used during the venturi calibration testing, the transducer uncertainty as determined by Eq. (7.51) is considered a fossilized systematic uncertainty. The uncertainty estimates for the venturi inlet and throat dimensions are based on the machining tolerances noted on the venturi manufacturer's data sheet. A systematic uncertainty of approximately 0.5% of reading is assigned to the kinematic viscosity to account for the uncertainty in the original experimental viscosity data used. (An argument could be made for a larger estimate of $2b_v$.)

The uncertainty in the discharge coefficient given by Eq. (7.54) was modeled as a function of Reynolds number with a second-order regression,

$$U_{C_d\text{-regress}}(\text{Re}) = 1.38 \times 10^{-12}\text{Re}^2 - 3.46 \times 10^{-7}\text{Re} + 0.0333 \qquad (7.57)$$

and is shown as the dashed lines along with the discharge coefficient data and regression model in Figure 7.5.

Also shown on Figure 7.5 are the uncertainty intervals for the individual (Re, C_d) data points. The discharge coefficient uncertainty for each data point was determined by applying the uncertainty propagation equations to Eq. (7.44) to obtain

$$U_{C_d}^2 = 2^2 \left[\left(\frac{\partial C_d}{\partial Q_{\text{std}}}\right)^2 s_{Q_{\text{std}}}^2 + \left(\frac{\partial C_d}{\partial Q_{\text{std}}}\right)^2 b_{Q_{\text{std}}}^2 + \left(\frac{\partial C_d}{\partial h}\right)^2 b_h^2 \right.$$

$$\left. + \left(\frac{\partial C_d}{\partial d}\right)^2 b_d^2 + \left(\frac{\partial C_d}{\partial D}\right)^2 b_D^2 \right] \qquad (7.58)$$

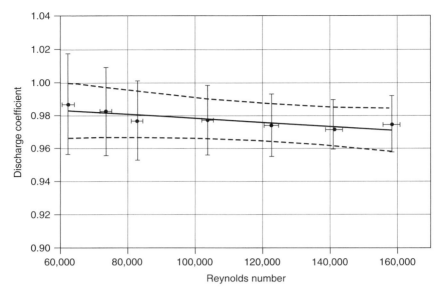

Figure 7.5   Discharge coefficient versus Reynolds number. (From Refs. 1 and 2.)

and the Reynolds number uncertainty was similarly determined as

$$U_{\mathrm{Re}}^2 = 2^2 \left[\left(\frac{\partial \mathrm{Re}}{\partial Q_{\mathrm{std}}}\right)^2 s_{Q_{\mathrm{std}}}^2 + \left(\frac{\partial \mathrm{Re}}{\partial Q_{\mathrm{std}}}\right)^2 b_{Q_{\mathrm{std}}}^2 + \left(\frac{\partial \mathrm{Re}}{\partial d}\right)^2 b_d^2 + \left(\frac{\partial \mathrm{Re}}{\partial v}\right)^2 b_v^2 \right]$$

$$(7.59)$$

It is observed in Figure 7.5 that the discharge coefficient uncertainty for individual data points is larger than the uncertainty associated with the discharge coefficient regression model at the same Re. This is reasonable, since the regression uncertainty interval at a given Re point includes information from the N data points used in the regression, while the uncertainty interval for a single data point at that Re includes information from that point only.

The uncertainty in C_d [given by Eq. (7.57) or Eq. (7.58)] is not as useful as supposed previously. When C_d is used in Eq. (7.47) or a_0 and a_1 are used in Eq. (7.48) to determine Q in an experiment, U_{c_d} or U_{a_0} and U_{a_1} contain(s) components that are correlated with the systematic errors in the values of d, D, and possibly v_{new} and h_{new} in Eq. (7.47) or Eq. (7.48). These additional correlation terms are difficult (or impossible) to formulate correctly, and the authors recommend the approach [1, 2] described in the following section.

7-7.4 Flow Rate and Its Uncertainty in a Test

As stated in the preceding section, it is difficult to propagate properly the discharge coefficient uncertainty to obtain the uncertainty associated with a flow-rate determination because of correlated systematic errors. In this section we demonstrate how the uncertainty associated with the venturi flowmeter calibration can be determined and how it can be reported in a form that properly expresses the uncertainty associated with the flow-rate determination.

When the venturi flowmeter is used in a test, the flow rate is calculated using Eq. (7.47) or (7.48). A seemingly straightforward method to determine the flow-rate uncertainty would be to apply the uncertainty propagation equations to Eq. (7.47) to obtain the uncertainty expression

$$U_Q^2 = 2^2 \left[\left(\frac{\partial Q}{\partial C_d} \right)^2 b_{C_d}^2 + \left(\frac{\partial Q}{\partial d} \right)^2 b_d^2 + \left(\frac{\partial Q}{\partial D} \right)^2 b_D^2 + \left(\frac{\partial Q}{\partial v_{\text{new}}} \right)^2 b_{v_{\text{new}}}^2 \right.$$

$$+ \left(\frac{\partial Q}{\partial h_{\text{new}}} \right)^2 (b_{h_{\text{new}}}^2 + s_{h_{\text{new}}}^2)$$

$$+ \left. \left\{ \begin{array}{c} \text{correlation terms between } C_d \text{ and } d, C_d \text{ and } D, \\ C_d \text{ and } h_{\text{new}}, \text{ and } C_d \text{ and } v_{\text{new}} \end{array} \right\} \right] \qquad (7.60)$$

or an equivalent expression with a_0 and a_1 appearing instead of C_d if Eq. (7.48) is used. The correlation terms are usually not considered in analyses estimating the contribution of the discharge coefficient uncertainty to the overall flow-rate uncertainty. Even if one wanted to consider them, their evaluation is difficult, if not impossible. The correlated systematic error effects must be taken into account, however, since some of the variables in Eq. (7.47) or Eq. (7.48) are the same variables in the $(\text{Re}(i), C_d(i))$ data set used in finding the discharge coefficient regression model [Eq. (7.46) or Eq. (7.52)]. The same values for the throat and inlet diameters d and D are used in Eq. (7.47) or (7.48) as were used in the determination of the discharge coefficient regression. If the same differential pressure transducer is used in the test to determine the flow rate as was used during the calibration or if a different transducer that has common systematic error sources with the calibration transducer is used to measure the differential pressure, appropriate correlation terms must be included. Similarly, if the value of kinematic viscosity has common systematic error sources with the values used during the discharge coefficient determination, the appropriate correlated systematic error terms must be included.

Recognizing that the flow rate is the result desired from the venturi during a test and that it is a function of the variables used to obtain the discharge coefficient regression as well as the variables measured in the test, the expression for Q can be represented functionally as

$$Q = f\{C_d(Q_{\text{std}}(i), h(i), v(i), d, D), h_{\text{new}}, v_{\text{new}}, d, D\} \qquad (7.61)$$

where C_d is the regression expression, Eq. (7.46) or (7.52), but with the regression coefficients replaced by expressions similar to those given by Eqs. (7.14) and (7.15). The uncertainty propagation equations can be applied directly to Eq. (7.61) to produce an expression for $U_{Q\text{-regress}}$.

The equation for $U_{Q\text{-regress}}$ can be written with the information known at the time of the venturi calibration and prior to the test; however, one must be careful to consider properly what systematic and random uncertainty information will only be known after the test has been conducted. If the same differential pressure transducer is used in the test as was used in the calibration and if the systematic uncertainty for the test value of kinematic viscosity is known, the only uncertainty not known prior to the test is the random uncertainty for the differential pressure measurements. Thus $U_{Q\text{-regress}}$ can be calculated as

$$
\begin{aligned}
U_{Q\text{-regress}}^2 = 2^2 \Bigg[& \sum_{i=1}^{N} \left(\frac{\partial Q}{\partial Q_{\text{std}(i)}} \right)^2 s_{Q_{\text{std}(i)}}^2 + \sum_{i=1}^{N} \left(\frac{\partial Q}{\partial h(i)} \right)^2 s_{h(i)}^2 \\
& + \sum_{i=1}^{N} \left(\frac{\partial Q}{\partial Q_{\text{std}(i)}} \right)^2 b_{Q_{\text{std}(i)}}^2 + 2 \sum_{i=1}^{N-1} \sum_{k=i+1}^{N} \left(\frac{\partial Q}{\partial Q_{\text{std}(i)}} \right) \\
& \times \left(\frac{\partial Q}{\partial Q_{\text{std}(k)}} \right) b_{Q_{\text{std}(i)}Q_{\text{std}(k)}} + \sum_{i=1}^{N} \left(\frac{\partial Q}{\partial h(i)} \right)^2 b_{h(i)}^2 \\
& + 2 \sum_{i=1}^{N-1} \sum_{k=i+1}^{N} \left(\frac{\partial Q}{\partial h(i)} \right) \left(\frac{\partial Q}{\partial h(k)} \right) b_{h(i)h(k)} + \sum_{i=1}^{N} \left(\frac{\partial Q}{\partial v(i)} \right)^2 b_{v(i)}^2 \\
& + 2 \sum_{i=1}^{N-1} \sum_{k=i+1}^{N} \left(\frac{\partial Q}{\partial v(i)} \right) \left(\frac{\partial Q}{\partial v(k)} \right) b_{v(i)v(k)} + \left(\frac{\partial Q}{\partial d} \right)^2 b_d^2 \\
& + \left(\frac{\partial Q}{\partial D} \right)^2 b_D^2 + \left(\frac{\partial Q}{\partial h_{\text{new}}} \right)^2 b_{h_{\text{new}}}^2 + 2 \sum_{i=1}^{N} \left(\frac{\partial Q}{\partial h(i)} \right) \\
& \times \left(\frac{\partial Q}{\partial h_{\text{new}}} \right) b_{h(i)h_{\text{new}}} + \left(\frac{\partial Q}{\partial v_{\text{new}}} \right)^2 b_{v_{\text{new}}}^2 + 2 \sum_{i=1}^{N} \left(\frac{\partial Q}{\partial v(i)} \right) \\
& \times \left(\frac{\partial Q}{\partial v_{\text{new}}} \right) b_{v(i)v_{\text{new}}} \Bigg]
\end{aligned}
\tag{7.62}
$$

where

$$
U_Q^2 = U_{Q\text{-regress}}^2 + 2^2 \left(\frac{\partial Q}{\partial h_{\text{new}}} \right)^2 s_{h_{\text{new}}}^2
\tag{7.63}
$$

and

$$
s_{h_{\text{new}}}^2 = \left(\frac{\partial h}{\partial V_{0,\text{new}}} \right)^2 s_{V_{0,\text{new}}}^2
\tag{7.64}
$$

The $b_{h(i)}$ and the $b_{h_{\text{new}}}$ in Eq. (7.62) are given by Eq. (7.51) as $b = U/2$. The correlation terms for h and Q_{std} are handled as discussed previously [Eq. (7.55)].

The first 10 terms on the RHS of Eq. (7.62) propagate the uncertainties associated with the venturi calibration into the flow-rate uncertainty. The remaining four terms of Eq. (7.62) propagate the uncertainties associated with the "new" variables when Eq. (7.48) is used to determine Q.

The effects of b_d and b_D in Eq. (7.62) deserve discussion. If the same values of d and D are used in the calibration and in Eq. (7.48), the $\partial Q/\partial d$ and $\partial Q/\partial D$ factors in terms 9 and 10 on the RHS of Eq. (7.62) will be zero. Thus the uncertainties in d and D contribute nothing to the uncertainty in Q—they have been "calibrated out." This makes sense if one realizes that using a value of d that is too large by a factor of 2 would not change anything as long as exactly the same value is used in the calibration and in determining Q using Eq. (7.48).

If the throat diameter d of the venturi changes (perhaps as a result of abrasion or erosion) and a different value of d is used in Eq. (7.48) than in the calibration, that d should be treated as a different variable (perhaps labeled d_{new}) than the d used in the calibration. In that case the $\partial Q/\partial d$ factor in term 9 on the RHS of Eq. (7.62) will no longer be zero, and an additional term, $[(\partial Q/\partial d_{new})^2(b_{d_{new}})^2]$, must be added to Eq. (7.62).

The kinematic viscosity terms in Eq. (7.62) also deserve discussion. If the same fluid is used in the calibration as in the test and if it has a constant kinematic viscosity and systematic uncertainty, the uncertainty in the kinematic viscosity has no effect on the uncertainty in Q. For this case, the four terms for kinematic viscosity in Eq. (7.62) add to zero.

The uncertainty $U_{Q\text{-regress}}$ determined after the calibration, as calculated using Eq. (7.62) and as a function of Reynolds number, is shown in Figure 7.6. A regression was performed using the set of (Re, $U_{Q\text{-regress}}$) points to give

$$U_{Q\text{-regress}} = 5 \times 10^{-16}\text{Re}_{new}^3 - 1 \times 10^{-10}\text{Re}_{new}^2$$
$$+ 2 \times 10^{-5}\text{Re}_{new} - 0.1162 \tag{7.65}$$

where $U_{Q\text{-regress}}$ is in gpm.

If the pressure transducer used during the test is not the same pressure transducer as was used during the calibration and does not share any common systematic error sources, the systematic uncertainty term for the new transducer and the $b_{h(i)h_{new}}$ term should be set equal to zero in Eq. (7.62), producing a new expression similar to Eq. (7.65). For the case where the test fluid kinematic viscosity systematic uncertainty is known, the resulting expression for the flow-rate uncertainty would then be

$$U_Q^2 = U_{Q\text{-regress}}^2 + 2^2\left(\frac{\partial Q}{\partial h_{new}}\right)^2 s_{h_{new}}^2 + 2^2\left(\frac{\partial Q}{\partial h_{new}}\right)^2 b_{h_{new}}^2 \tag{7.66}$$

The regression giving $U_{Q\text{-regress}}$ [Eq. (7.65) or a similar one] should be reported along with Eq. (7.63) or (7.66), as appropriate.

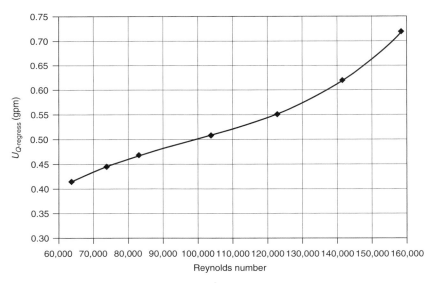

Figure 7.6   Uncertainty in venturi flow rate as a function of Reynolds Number. (From Refs. 1 and 2.)

7-8 MULTIPLE LINEAR REGRESSION

The basic linear least-squares analysis presented in Section 7-2 can be extended to the case in which the result of the experiment is a function of more than one variable. This dependence on more than one variable is typically the case, but we often choose to report the results as a function of one of the variables with the others held fixed. For instance, we might determine a heat transfer relationship for Nusselt number Nu versus Reynolds number Re for a specific geometry (length-to-diameter ratio, L/D) and fluid. If we had more data for several other L/D ratios, we could consider using a multiple linear regression to obtain a correlation equation to represent all the results.

Consider an experimental result R that is a function of three variables, X, Y, and Z. If we assume that R varies linearly with respect to X, Y, and Z, we can write the expression

$$R_0 = a_1 X + a_2 Y + a_3 Z + a_4 \tag{7.67}$$

where R_0 is taken as the "optimum" R value to represent the data for a given X, Y, and Z. Just as in Section 7-2, we wish to find the values of a_1, a_2, a_3, and a_4 that minimize

$$\eta = \sum_{i=1}^{N} (R_i - a_1 X_i - a_2 Y_i - a_3 Z_i - a_4)^2 \tag{7.68}$$

where X_i, Y_i, and Z_i are the N data values.

By taking the derivatives of η with respect to the constants a_i and setting these expressions equal to zero, we obtain four equations with four unknowns:

$$a_1 \sum X_i^2 + a_2 \sum X_i Y_i + a_3 \sum X_i Z_i + a_4 \sum X_i = \sum R_i X_i \qquad (7.69)$$

$$a_1 \sum X_i Y_i + a_2 \sum Y_i^2 + a_3 \sum Y_i Z_i + a_4 \sum Y_i = \sum R_i Y_i \qquad (7.70)$$

$$a_1 \sum X_i Z_i + a_2 \sum Y_i Z_i + a_3 \sum Z_i^2 + a_4 \sum Z_i = \sum R_i Z_i \qquad (7.71)$$

$$a_1 \sum X_i + a_2 \sum Y_i + a_3 \sum Z_i + a_4 N = \sum R_i \qquad (7.72)$$

where the summations run from $i = 1$ to $i = N$. In matrix notation, Eqs. (7.69) to (7.72) become

$$\begin{bmatrix} \sum X_i^2 & \sum X_i Y_i & \sum X_i Z_i & \sum X_i \\ \sum X_i Y_i & \sum Y_i^2 & \sum Y_i Z_i & \sum Y_i \\ \sum X_i Z_i & \sum Y_i Z_i & \sum Z_i^2 & \sum Z_i \\ \sum X_i & \sum Y_i & \sum Z_i & N \end{bmatrix} \begin{bmatrix} a_1 \\ a_2 \\ a_3 \\ a_4 \end{bmatrix} = \begin{bmatrix} \sum R_i X_i \\ \sum R_i Y_i \\ \sum R_i Z_i \\ \sum R_i \end{bmatrix} \qquad (7.73)$$

or

$$[C][a] = [R] \qquad (7.74)$$

where $[C]$, $[a]$, and $[R]$ represent the matrices in Eq. (7.73). The solution for $[a]$ can then be expressed as

$$[a] = [C]^{-1}[R] \qquad (7.75)$$

where $[C]^{-1}$ is the inverse of the coefficient matrix.

The curve fit that results from this procedure is

$$R = a_1 X + a_2 Y + a_3 Z + a_4 \qquad (7.76)$$

If the result is only a function of two variables (X, Y), the terms containing Z in Eqs. (7.67) to (7.70) will be zero, and the coefficient matrix in Eq. (7.73) will be a 3×3 matrix. This procedure can also be extended for the case in which R is a function of more than three variables.

The uncertainty that should be associated with a value R calculated using the curve fit is once again determined using the general approach in Section 7-4. The curve-fit equation should be plotted with the data points to observe how well the equation represents the data.

The condition given in Eq. (7.67) that the result R is a linear function of the independent variables may seem too restrictive at first. However, the variables X, Y, and Z do not have to be linear functions themselves, as the concept of rectification allows us to convert a nonlinear data representation equation into a linear form.

Consider the following example. Heat transfer data for external flow conditions are usually expressed as

$$St = k\,Re^a\,Pr^b \tag{7.77}$$

where St is the Stanton number, Re is the Reynolds number, Pr is the Prandtl number, and k is a constant. If we take the logarithm of both sides of Eq. (7.77), we obtain

$$\log St = \log k + a \log Re + b \log Pr \tag{7.78}$$

or in the notation and format of Eq. (7.67),

$$\log St = a_1 \log Re + a_2 \log Pr + a_3 \tag{7.79}$$

This form of representation with log St as the result and log Re and log Pr as the independent variables is in the required linear format. The curve-fit constants a_i can be obtained using the technique given in Eq. (7.75).

REFERENCES

1. Brown, K. K., "Assessment of the Experimental Uncertainty Associated with Regressions," Ph.D. dissertation, University of Alabama in Huntsville, Huntsville, AL, 1996.
2. Brown, K. K., Coleman, H. W., and Steele, W. G., "A Methodology for Determining the Experimental Uncertainty in Regressions," *Journal of Fluids Engineering*, Vol. 120, No. 3, 1998, pp. 445–456.
3. Schenck, H., *Theories of Engineering Experimentation*, 3rd ed., McGraw-Hill, New York, 1979.
4. Montgomery, D. C., and Peck, E. A., *Introduction to Linear Regression Analysis*, 2nd ed., Wiley, New York, 1992.
5. American National Standards Institute/American Society of Mechanical Engineers (ASME), *Test Uncertainty*, PTC 19.1-2005, ASME, New York, 2006.
6. Press, W. H., Flannery, B. P., Teukolsky, S. A., and Vetterling, W. T., *Numerical Recipes: The Art of Scientific Computing*, Cambridge University Press, Cambridge, 1986.
7. Bean, H. S. (Ed.), *Fluid Meters: Their Theory and Application*, 6th ed., American Society of Mechanical Engineers, New York, 1971.

PROBLEMS

7.1 Determine a regression equation for the following set of data. Specify the 95% random uncertainty associated with the use of this equation to determine a value of Y at $X = 3.2$, $X = 5.0$, and $X = 7.0$.

Y	2.4	3.0	3.5	4.5	4.9	5.6	6.8	7.3
X	2.0	3.0	4.5	5.3	6.5	7.8	8.5	10.1

What are the 95% random uncertainties for the slope m and intercept c?

7.2 In the calibration of an orifice plate for airflow measurements, the following volumetric flow rate (Q) versus pressure drop (ΔP) data were obtained:

Q (ft^3/min)	130	304	400	488	554	610
ΔP (psi)	5	25	45	65	85	105

It is known that an appropriate regression equation for this situation is $Q = C(\Delta P)^{1/2}$, where Q is the flow in ft^3/min, ΔP is the orifice pressure drop in psi, and C is the calibration factor. Plot these data in rectified coordinates and determine C. What is the systematic standard uncertainty in C if the systematic standard uncertainty for the flowmeter used to obtain Q is 0.5% of the reading and the systematic standard uncertainty for the ΔP cell is 0.5% of the reading?

7.3 The expected response of a first-order instrument to a step change in input follows the relationship

$$R = 1 - e^{-t/\tau}$$

Data for response versus time is

R	0.39	0.63	0.78	0.86	0.92
t (s)	1	2	3	4	5

Plot these data in the appropriate rectified coordinate system and find the regression equation for τ. Include the curve-fit uncertainty bands when $b_R = 0.025$, $b_t = 0.05$ s, $s_R = 0.01$, and $s_t = 0.025$ s.

7.4 The drag on a cylinder for turbulent crossflow in air is to be expressed as a drag coefficient C_D,

$$C_D = \frac{F_D}{A_f(\rho V^2/2)}$$

versus Reynolds number Re,

$$\text{Re} = \frac{\rho V D}{\mu}$$

where F_D is the force on the cylinder, A_f is the cylinder frontal (projected) area, ρ is the density of the air, V is the air free-stream velocity, D is the diameter of the cylinder, and μ is the dynamic viscosity of the air. A cylinder

6 cm. in diameter and 60 cm. long is attached to a load cell and placed in a wind tunnel. The load cell measures the force F_D on the cylinder. The velocity V is adjusted with a variable speed fan and is measured with a pitot probe.

The following data have been taken for force and velocity:

Data Point	F_D (N)	V (m/s)
1	35.2	53
2	46.2	59
3	62.9	67
4	83.5	75
5	109.7	84
6	144.0	94
7	188.9	105

The 95% systematic uncertainty in the force is primarily from the load cell calibration and has a value of 0.7 N. The 95% systematic uncertainty in the velocity is primarily from the differential pressure (DP) cell uncertainty and is 0.5 m/s. The random uncertainty for the two measurements is considered negligible.

The diameter of the cylinder is determined with a micrometer, and the total (fossilized) 95% systematic uncertainty is 0.05 cm. The length is measured with a rule, and the 95% systematic uncertainty is 0.2 cm. The air density is 1.161 kg/m^3 $\pm$ 1%(95%), and the dynamic viscosity is 184.6 $\times$ 10^{-7} Ns/m^2 $\pm$ 2%(95%).

(a) Find the appropriate regression expression for these data. The appropriate rectified coordinate system for this range of values is drag coefficient versus log Re.

(b) For the case where this regression expression is used for initial design purposes with $U_{Re(new)} = 0$, determine the uncertainty statement for the curve fit.

(c) For the case where Re(new) is a design Reynolds number for a different test and where it has a systematic standard uncertainty of $b_{Re(new)}$, what is the expression for the uncertainty in the drag coefficient determined from the curve fit?

APPENDIX A

USEFUL STATISTICS

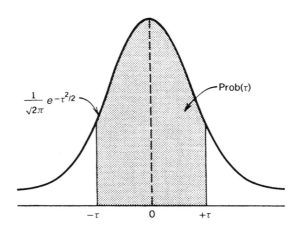

Figure A.1 Graphic representatsion of the two-tailed Gaussian probability.

Table A.1   Tabulation of Two-Tailed Gaussian Probabilities

τ	Prob (τ)	τ	Prob (τ)	τ	Prob (τ)	τ	Prob (τ)
0.00	0.0000	1.00	0.6827	2.00	0.9545	3.00	0.9973002
0.02	0.0160	1.02	0.6923	2.02	0.9566	3.05	0.9977115
0.04	0.0319	1.04	0.7017	2.04	0.9586	3.10	0.9980647
0.06	0.0478	1.06	0.7109	2.06	0.9606	3.15	0.9983672
0.08	0.0638	1.08	0.7199	2.08	0.9625	3.20	0.9986257
0.10	0.0797	1.10	0.7287	2.10	0.9643	3.25	0.9988459
0.12	0.0955	1.12	0.7373	2.12	0.9660	3.30	0.9990331
0.14	0.1113	1.14	0.7457	2.14	0.9676	3.35	0.9991918
0.16	0.1271	1.16	0.7540	2.16	0.9692	3.40	0.9993261
0.18	0.1428	1.18	0.7620	2.18	0.9707	3.45	0.9994394
0.20	0.1585	1.20	0.7699	2.20	0.9722	3.50	0.9995347
0.22	0.1741	1.22	0.7775	2.22	0.9736	3.55	0.9996147
0.24	0.1897	1.24	0.7850	2.24	0.9749	3.60	0.9996817
0.26	0.2051	1.26	0.7923	2.26	0.9762	3.65	0.9997377
0.28	0.2205	1.28	0.7995	2.28	0.9774	3.70	0.9997843
0.30	0.2358	1.30	0.8064	2.30	0.9786	3.75	0.9998231
0.32	0.2510	1.32	0.8132	2.32	0.9797	3.80	0.9998552
0.34	0.2661	1.34	0.8198	2.34	0.9807	3.85	0.9998818
0.36	0.2812	1.36	0.8262	2.36	0.9817	3.90	0.9999037
0.38	0.2961	1.38	0.8324	2.38	0.9827	3.95	0.9999218
0.40	0.3108	1.40	0.8385	2.40	0.9836	4.00	0.9999366
0.42	0.3255	1.42	0.8444	2.42	0.9845	4.05	0.9999487
0.44	0.3401	1.44	0.8501	2.44	0.9853	4.10	0.9999586
0.46	0.3545	1.46	0.8557	2.46	0.9861	4.15	0.9999667
0.48	0.3688	1.48	0.8611	2.48	0.9869	4.20	0.9999732
0.50	0.3829	1.50	0.8664	2.50	0.9876	4.25	0.9999786
0.52	0.3969	1.52	0.8715	2.52	0.9883	4.30	0.9999829
0.54	0.4108	1.54	0.8764	2.54	0.9889	4.35	0.9999863
0.56	0.4245	1.56	0.8812	2.56	0.9895	4.40	0.9999891
0.58	0.4381	1.58	0.8859	2.58	0.9901	4.45	0.9999911
0.60	0.4515	1.60	0.8904	2.60	0.9907	4.50	0.9999931
0.62	0.4647	1.62	0.8948	2.62	0.9912	4.55	0.9999946
0.64	0.4778	1.64	0.8990	2.64	0.9917	4.60	0.9999957
0.66	0.4907	1.66	0.9031	2.66	0.9922	4.65	0.9999966
0.68	0.5035	1.68	0.9070	2.68	0.9926	4.70	0.9999973
0.70	0.5161	1.70	0.9109	2.70	0.9931	4.75	0.9999979
0.72	0.5285	1.72	0.9146	2.72	0.9935	4.80	0.9999984
0.74	0.5407	1.74	0.9181	2.74	0.9939	4.85	0.9999987
0.76	0.5527	1.76	0.9216	2.76	0.9942	4.90	0.9999990
0.78	0.5646	1.78	0.9249	2.78	0.9946	4.95	0.9999992
0.80	0.5763	1.80	0.9281	2.80	0.9949	5.00	0.9999994
0.82	0.5878	1.82	0.9312	2.82	0.9952		
0.84	0.5991	1.84	0.9342	2.84	0.9955		
0.86	0.6102	1.86	0.9371	2.86	0.9958		
0.88	0.6211	1.88	0.9399	2.88	0.9960		
0.90	0.6319	1.90	0.9426	2.90	0.9963		
0.92	0.6424	1.92	0.9451	2.92	0.9965		
0.94	0.6528	1.94	0.9476	2.94	0.9967		
0.96	0.6629	1.96	0.9500	2.96	0.9969		
0.98	0.6729	1.98	0.9523	2.98	0.9971		

Table A.2 The t Distribution[a]

ν	C				
	0.900	0.950	0.990	0.995	0.999
1	6.314	12.706	63.657	127.321	636.619
2	2.920	4.303	9.925	14.089	31.598
3	2.353	3.182	5.841	7.453	12.924
4	2.132	2.776	4.604	5.598	8.610
5	2.015	2.571	4.032	4.773	6.869
6	1.943	2.447	3.707	4.317	5.959
7	1.895	2.365	3.499	4.029	5.408
8	1.860	2.306	3.355	3.833	5.041
9	1.833	2.262	3.250	3.690	4.781
10	1.812	2.228	3.169	3.581	4.587
11	1.796	2.201	3.106	3.497	4.437
12	1.782	2.179	3.055	3.428	4.318
13	1.771	2.160	3.012	3.372	4.221
14	1.761	2.145	2.977	3.326	4.140
15	1.753	2.131	2.947	3.286	4.073
16	1.746	2.120	2.921	3.252	4.015
17	1.740	2.110	2.898	3.223	3.965
18	1.734	2.101	2.878	3.197	3.922
19	1.729	2.093	2.861	3.174	3.883
20	1.725	2.086	2.845	3.153	3.850
21	1.721	2.080	2.831	3.135	3.819
22	1.717	2.074	2.819	3.119	3.792
23	1.714	2.069	2.807	3.104	3.768
24	1.711	2.064	2.797	3.090	3.745
25	1.708	2.060	2.787	3.078	3.725
26	1.706	2.056	2.779	3.067	3.707
27	1.703	2.052	2.771	3.057	3.690
28	1.701	2.048	2.763	3.047	3.674
29	1.699	2.045	2.756	3.038	3.659
30	1.697	2.042	2.750	3.030	3.646
40	1.684	2.021	2.704	2.971	3.551
60	1.671	2.000	2.660	2.915	3.460
120	1.658	1.980	2.617	2.860	3.373
∞	1.645	1.960	2.576	2.807	3.291

[a]Given are the values of t for a confidence level C and number of degrees of freedom ν.

Table A.3　Factors for Tolerance Interval

N	90% Confidence That Percentage of Population between Limits Is:			95% Confidence That Percentage of Population between Limits Is:			99% Confidence That Percentage of Population between Limits Is:		
	90%	95%	99%	90%	95%	99%	90%	95%	99%
2	15.98	18.80	24.17	32.02	37.67	48.43	160.2	188.5	242.3
3	5.847	6.919	8.974	8.380	9.916	12.86	18.93	22.40	29.06
4	4.166	4.943	6.440	5.369	6.370	8.299	9.398	11.15	14.53
5	3.494	4.152	5.423	4.275	5.079	6.634	6.612	7.855	10.26
6	3.131	3.723	4.870	3.712	4.414	5.775	5.337	6.345	8.301
7	2.902	3.452	4.521	3.369	4.007	5.248	4.613	5.488	7.187
8	2.743	3.264	4.278	3.136	3.732	4.891	4.147	4.936	6.468
9	2.626	3.125	4.098	2.967	3.532	4.631	3.822	4.550	5.966
10	2.535	3.018	3.959	2.839	3.379	4.433	3.582	4.265	5.594
11	2.463	2.933	3.849	2.737	3.259	4.277	3.397	4.045	5.308
12	2.404	2.863	3.758	2.655	3.162	4.150	3.250	3.870	5.079
13	2.355	2.805	3.682	2.587	3.081	4.044	3.130	3.727	4.893
14	2.314	2.756	3.618	2.529	3.012	3.955	3.029	3.608	4.737
15	2.278	2.713	3.562	2.480	2.954	3.878	2.945	3.507	4.605
16	2.246	2.676	3.514	2.437	2.903	3.812	2.872	3.421	4.492
17	2.219	2.643	3.471	2.400	2.858	3.754	2.808	3.345	4.393
18	2.194	2.614	3.433	2.366	2.819	3.702	2.753	3.279	4.307
19	2.172	2.588	3.399	2.337	2.784	3.656	2.703	3.221	4.230
20	2.152	2.564	3.368	2.310	2.752	3.615	2.659	3.168	4.161
21	2.135	2.543	3.340	2.286	2.723	3.577	2.620	3.121	4.100
22	2.118	2.524	3.315	2.264	2.697	3.543	2.584	3.078	4.044
23	2.103	2.506	3.292	2.244	2.673	3.512	2.551	3.040	3.993
24	2.089	2.489	3.270	2.225	2.651	3.483	2.522	3.004	3.947

25	2.077	2.474	3.251	2.208	2.631	3.457	2.494	2.972	3.904
26	2.065	2.460	3.232	2.193	2.612	3.432	2.469	2.941	3.865
27	2.054	2.447	3.215	2.178	2.595	3.409	2.446	2.914	3.828
28	2.044	2.435	3.199	2.164	2.579	3.388	2.424	2.888	3.794
29	2.034	2.424	3.184	2.152	2.554	3.368	2.404	2.864	3.763
30	2.025	2.413	3.170	2.140	2.549	3.350	2.385	2.841	3.733
35	1.988	2.368	3.112	2.090	2.490	3.272	2.306	2.748	3.611
40	1.959	2.334	3.066	2.052	2.445	3.213	2.247	2.677	3.518
50	1.916	2.284	3.001	1.996	2.379	3.126	2.162	2.576	3.385
60	1.887	2.248	2.955	1.958	2.333	3.066	2.103	2.506	3.293
80	1.848	2.202	2.894	1.907	2.272	2.986	2.026	2.414	3.173
100	1.822	2.172	2.854	1.874	2.233	2.934	1.977	2.355	3.096
200	1.764	2.102	2.762	1.798	2.143	2.816	1.865	2.222	2.921
500	1.717	2.046	2.689	1.737	2.070	2.721	1.777	2.117	2.783
1000	1.695	2.019	2.654	1.709	2.036	2.676	1.736	2.068	2.718
∞	1.645	1.960	2.576	1.645	1.960	2.576	1.645	1.960	2.576

Source: D.C. Montgomery and G. C. Runger, *Applied Statistics and Probability for Engineers*, Wiley, New York. Copyright © 1994 John Wiley & Sons, Inc. Reprinted by permission of John Wiley & Sons, Inc.

Table A.4   Factors for Prediction Interval to Contain the Values of All of 1, 2, 5, 10, and 20 Future Observations at a Confidence Level of 95%

N	$c_{p,1}(N)$	$c_{p,2}(N)$	$c_{p,5}(N)$	$c_{p,10}(N)$	$c_{p,20}(N)$
4	3.56	4.41	5.56	6.41	7.21
5	3.04	3.70	4.58	5.23	5.85
6	2.78	3.33	4.08	4.63	5.16
7	2.62	3.11	3.77	4.26	4.74
8	2.51	2.97	3.57	4.02	4.46
9	2.43	2.86	3.43	3.85	4.26
10	2.37	2.79	3.32	3.72	4.10
11	2.33	2.72	3.24	3.62	3.98
12	2.29	2.68	3.17	3.53	3.89
15	2.22	2.57	3.03	3.36	3.69
20	2.14	2.48	2.90	3.21	3.50
25	2.10	2.43	2.83	3.12	3.40
30	2.08	2.39	2.78	3.06	3.33
40	2.05	2.35	2.73	2.99	3.25
60	2.02	2.31	2.67	2.93	3.17
∞	1.96	2.24	2.57	2.80	3.02

Source: G. Hahn, "Understanding Statistical Intervals," *Industrial Engineering*, December 1970. pp. 45–48.

APPENDIX B

TAYLOR SERIES METHOD (TSM) FOR UNCERTAINTY PROPAGATION[1]

In nearly all experiments, the measured values of different variables are combined using a data reduction equation (DRE) to form some desired result. A good example is the experimental determination of drag coefficient of a particular model configuration in a wind tunnel test. Defining drag coefficient as

$$C_D = \frac{2F_D}{\rho V^2 A} \tag{B.1}$$

one can envision that errors in the values of the variables on the right-hand side of Eq. (B.1) will cause errors in the experimental result C_D.

A more general representation of a data reduction equation is

$$r = r(X_1, X_2, \ldots, X_J) \tag{B.2}$$

where r is the experimental result determined from J measured variables X_i. Each of the measured variables contains systematic (bias) errors and random (precision) errors. These errors in the measured values then propagate through the data reduction equation, thereby generating the systematic and random errors in the experimental result, r. Our goal in uncertainty analysis is to determine the effects of these errors, which result in the random and systematic uncertainties in the result. In this appendix, a derivation of the equation describing uncertainty propagation is presented, comparisons with previously used equations and approaches are discussed, and, finally, the approximations leading to the method we recommend for most engineering applications are described.

---

[1] Appendix B is adapted from Ref. 1.

B-1 DERIVATION OF UNCERTAINTY PROPAGATION EQUATION

Rather than present the derivation for the case in which the result is a function of many variables, the simpler case in which the result is a function of only two variables is considered first. The expressions for the more general case will then be presented as extensions of the two-variable case.

Suppose that the data reduction equation is

$$r = r(x, y) \tag{B.3}$$

where the function is continuous and has continuous derivatives in the domain of interest. The situation is shown schematically in Figure B.1 for the kth set of measurements (x_k, y_k) which is used to determine r_k. Here β_{x_k} and ϵ_{x_k} are the systematic and random errors, respectively, in the kth measurement of x, with a similar convention for the errors in y and in r. Assume that the test instrumentation and/or apparatus is changed for each measurement so that different values of β_{x_k} and β_{y_k} will occur for each measurement. Therefore, the systematic errors and random errors will be random variables, so

$$x_k = x_{\text{true}} + \beta_{x_k} + \epsilon_{x_k} \tag{B.4}$$

$$y_k = y_{\text{true}} + \beta_{y_k} + \epsilon_{y_k} \tag{B.5}$$

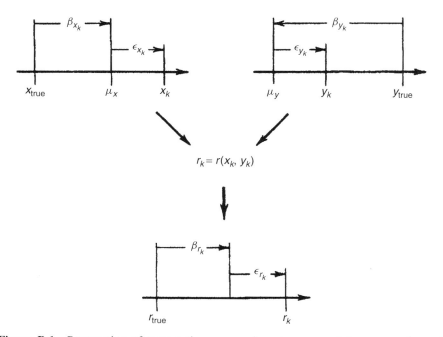

Figure B.1 Propagation of systematic errors and random errors into an experimental result.

Now approximate the function r in the DRE using a Taylor series expansion. Expanding to the general point r_k from r_{true} gives

$$r_k = r_{\text{true}} + \frac{\partial r}{\partial x}(x_k - x_{\text{true}}) + \frac{\partial r}{\partial y}(y_k - y_{\text{true}}) + R_2 \tag{B.6}$$

where R_2 is the remainder term and where the partial derivatives are evaluated at $(x_{\text{true}}, y_{\text{true}})$. Since the true values of x and y are unknown, an approximation is always introduced when the derivatives are evaluated at some measured values (x_k, y_k).

The remainder term has the form [2]

$$R_2 = \frac{1}{2!}\left[\frac{\partial^2 r}{\partial x^2}(x_k - x_{\text{true}})^2 + 2\frac{\partial^2 r}{\partial x \partial y}(x_k - x_{\text{true}})(y_k - y_{\text{true}}) + \frac{\partial^2 r}{\partial y^2}(y_k - y_{\text{true}})^2\right] \tag{B.7}$$

where the partial derivatives are evaluated at (ζ, χ), which is somewhere between (x_k, y_k) and $(x_{\text{true}}, y_{\text{true}})$. This term is usually assumed to be negligible, so it is useful to consider the conditions under which this assumption might be reasonable. The factors $x_k - x_{\text{true}}$ and $y_k - y_{\text{true}}$ are the total errors in x and y. If the derivatives are of reasonable magnitude and the total errors in x and y are small, then R_2, containing the squares of the errors, will approach zero more quickly than will the first-order terms. Also, if $r(x, y)$ is a linear function, the partial derivatives in Eq. (B.7) are identically zero (as is R_2).

Neglecting R_2, the expansion gives [taking r_{true} to the left-hand side (LHS)]

$$(r_k - r_{\text{true}}) = \frac{\partial r}{\partial x}(x_k - x_{\text{true}}) + \frac{\partial r}{\partial y}(y_k - y_{\text{true}}) \tag{B.8}$$

This expression relates the total error δ in the kth determination of the result r to the total errors in the measured variables and using the notation

$$\theta_x = \frac{\partial r}{\partial x} \tag{B.9}$$

can be written as

$$\delta_{r_k} = \theta_x(\beta_{x_k} + \epsilon_{x_k}) + \theta_y(\beta_{y_k} + \epsilon_{y_k}) \tag{B.10}$$

We are interested in obtaining a measure of the distribution of the δ_r values for (some large number) N determinations of the result r. The variance of the parent distribution is defined by

$$\sigma_{\delta_r}^2 = \lim(N \to \infty)\left[\frac{1}{N}\sum_{k=1}^{N}(\delta_{r_k})^2\right] \tag{B.11}$$

Substituting (B.10) into (B.11) but deferring taking the limit gives

$$\frac{1}{N}\sum_{k=1}^{N}(\delta_{r_k})^2 = \theta_x^2\frac{1}{N}\sum_{k=1}^{N}(\beta_{x_k})^2 + \theta_y^2\frac{1}{N}\sum_{k=1}^{N}(\beta_{y_k})^2 + 2\theta_x\theta_y\frac{1}{N}\sum_{k=1}^{N}\beta_{x_k}\beta_{y_k}$$

$$+ \theta_x^2\frac{1}{N}\sum_{k=1}^{N}(\epsilon_{x_k})^2 + \theta_y^2\frac{1}{N}\sum_{k=1}^{N}(\epsilon_{y_k})^2 + 2\theta_x\theta_y\frac{1}{N}\sum_{k=1}^{N}\epsilon_{x_k}\epsilon_{y_k}$$

$$+ 2\theta_x^2\frac{1}{N}\sum_{k=1}^{N}\beta_{x_k}\epsilon_{x_k} + 2\theta_y^2\frac{1}{N}\sum_{k=1}^{N}\beta_{y_k}\epsilon_{y_k}$$

$$+ 2\theta_x\theta_y\frac{1}{N}\sum_{k=1}^{N}\beta_{x_k}\epsilon_{y_k} + 2\theta_x\theta_y\frac{1}{N}\sum_{k=1}^{N}\beta_{y_k}\epsilon_{x_k} \tag{B.12}$$

Taking the limit as N approaches infinity and using definitions of variances similar to that in Eq. (B.11), we obtain

$$\sigma_{\delta_r}^2 = \theta_x^2\sigma_{\beta_x}^2 + \theta_y^2\sigma_{\beta_y}^2 + 2\theta_x\theta_y\sigma_{\beta_x\beta_y} + \theta_x^2\sigma_{\epsilon_x}^2 + \theta_y^2\sigma_{\epsilon_y}^2 + 2\theta_x\theta_y\sigma_{\epsilon_x\epsilon_y} \tag{B.13}$$

assuming that there are no systematic error/random error correlations so that in Eq. (B.12) the final four terms containing the $\beta\epsilon$ products are zero.

Since in reality we never know the σ values exactly, we must use *estimates* of them. Defining u_c^2 as an estimate of the variance of the distribution of total errors in the result, b^2 as an estimate of the variance of a systematic error distribution, and s^2 as an estimate of the variance of a random error distribution, we can write

$$u_c^2 = \theta_x^2 b_x^2 + \theta_y^2 b_y^2 + 2\theta_x\theta_y b_{xy} + \theta_x^2 s_x^2 + \theta_y^2 s_y^2 + 2\theta_x\theta_y s_{xy} \tag{B.14}$$

In Eq. (B.14), b_{xy} is an estimate of the covariance of the systematic errors in x and the systematic errors in y that is defined exactly by

$$\sigma_{\beta_x\beta_y} = \lim(N \to \infty)\left(\frac{1}{N}\sum_{k=1}^{N}\beta_{x_k}\beta_{y_k}\right) \tag{B.15}$$

Similarly, s_{xy} is an estimate of the covariance of the random errors in x and y. In keeping with the nomenclature of the ISO guide [3], u_c is called the *combined standard uncertainty*. For the more general case in which the experimental result is determined from Eq. (B.2), u_c is given by

$$u_c^2 = \sum_{i=1}^{J}\theta_i^2 b_i^2 + 2\sum_{i=1}^{J-1}\sum_{k=i+1}^{J}\theta_i\theta_k b_{ik} + \sum_{i=1}^{J}\theta_i^2 s_i^2 + 2\sum_{i=1}^{J-1}\sum_{k=i+1}^{J}\theta_i\theta_k s_{ik} \tag{B.16}$$

where b_i^2 is the estimate of the variance of the systematic error distribution of variable X_i, and so on. The derivation to this point was presented by the authors in Ref. 4.

No assumptions about type(s) of error distributions are made to obtain the preceding equation for u_c. To obtain an uncertainty U_r (termed the *expanded uncertainty* in the ISO guide) at some specified confidence level (95%, 99%, etc.), the combined standard uncertainty u_c must be multiplied by a coverage factor K,

$$U_r = K u_c \tag{B.17}$$

It is in choosing K that assumptions about the type(s) of the error distributions must be made.

An argument is presented in the ISO guide that the error distribution of the result r may often be considered Gaussian because of the central limit theorem, even if the error distributions of the X_i are not normal. In fact, the same argument can be made for approximate normality of the error distributions of the X_i since the errors typically are composed of a combination of errors from a number of elemental sources.

If it is assumed that the error distribution of the result r is normal, the value of K for C percent coverage corresponds to the C percent confidence level t value from the t distribution (Appendix A), so that

$$U_r^2 = t^2 \left[\sum_{i=1}^{J} \theta_i^2 b_i^2 + 2 \sum_{i=1}^{J-1} \sum_{k=i+1}^{J} \theta_i \theta_k b_{ik} \right.$$
$$\left. + \sum_{i=1}^{J} \theta_i^2 s_i^2 + 2 \sum_{i=1}^{J-1} \sum_{k=i+1}^{J} \theta_i \theta_k s_{ik} \right] \tag{B.18}$$

The effective number of degrees of freedom ν_r for determining the t value is given (approximately) by the Welch–Satterthwaite formula as [3]

$$\nu_r = \frac{\left[\sum_{i=1}^{J} (\theta_i^2 s_i^2 + \theta_i^2 b_i^2) \right]^2}{\sum_{i=1}^{J} \{ [(\theta_i s_i)^4 / \nu_{s_i}] + [(\theta_i b_i)^4 / \nu_{b_i}] \}} \tag{B.19}$$

where the ν_{s_i} are the number of degrees of freedom associated with the s_i and the ν_{b_i} are the number of degrees of freedom to associate with the b_i.

If an s_i has been determined from N_i readings of X_i taken over an appropriate interval, the number of degrees of freedom is given by

$$\nu_{s_i} = N_i - 1 \tag{B.20}$$

For the number of degrees of freedom ν_{b_i} to associate with a nonstatistical estimate of b_i, it is suggested in the ISO guide that one might use the approximation

$$\nu_{b_i} \approx \frac{1}{2} \left(\frac{\Delta b_i}{b_i} \right)^{-2} \tag{B.21}$$

where the quantity in parentheses is the relative uncertainty of b_i. For example, if one thought that the estimate of b_i was reliable to within $\pm 25\%$,

$$v_{b_i} \approx \tfrac{1}{2}(0.25)^{-2} \approx 8 \tag{B.22}$$

Consideration of the 95% confidence t values in Table A.2 reveals that the value of t approaches 2.0 (approximately) as the number of degrees of freedom increases. We thus face the somewhat paradoxical situation that as we have more information (v_r increases), we can take t equal to 2.0 and do not have to deal with Eq. (B.19), but for the cases in which we have little information (v_r small), we need to make the more detailed estimates required by Eq. (B.19). In Section B-3 we examine the assumptions required to discard Eq. (B.19) and to simply use $t = 2$ (for 95% confidence), but first in Section B-2 we compare the derived approach with those published earlier.

B-2 COMPARISON WITH PREVIOUS APPROACHES

The purpose of this section is to give an historical perspective of the development of uncertainty analysis methodology.

B-2.1 Abernethy et al. Approach

An approach that was widely used in the 1970s and 1980s was the U_{RSS}, U_{ADD} technique formulated by Abernethy and co-workers [5] and used in Refs. [6] and [7] and other SAE, ISA, JANNAF, NRC, USAF, NATO, and ISO standards documents [5]. According to Abernethy et al.,

$$U_{\mathrm{RSS}} = [B_r^2 + (ts_r)^2]^{1/2} \tag{B.23}$$

for a 95% confidence estimate and

$$U_{\mathrm{ADD}} = B_r + ts_r \tag{B.24}$$

for a 99% confidence estimate, where B_r is given by

$$B_r = \left[\sum_{i=1}^{J} (\theta_i^2 B_i^2) \right]^{1/2} \tag{B.25}$$

and B_r and the B_i values are 95% confidence systematic uncertainty (bias limit) estimates, and

$$s_r = \left[\sum_{i=1}^{J} (\theta_i^2 s_i^2) \right]^{1/2} \tag{B.26}$$

where t is the 95% confidence t value from the t distribution for ν_r degrees of freedom given by

$$\nu_r = \frac{(\theta_i s_i)^4}{\sum_{i=1}^{J} [(\theta_i s_i)^4 / \nu_{s_i}]} \tag{B.27}$$

Consideration of these expressions in the context of the derivation presented in the preceding section shows that they cannot be justified on a rigorous basis. The U_{ADD} approach has always been advanced on the basis of ad hoc arguments and with results from a few Monte Carlo simulations, but (as argued in the ISO guide [3]) for a 99% confidence level, the t value appropriate for 99% confidence should be used as the value of K in Eq. (B.17) to obtain a 99% confidence estimate for an assumed Gaussian distribution. The Abernethy et al. approaches also ignore the possibility of correlated systematic error effects [taken into account in the b_{ik} covariance terms in Eq. (B.18)], although Ref. 6 does consider this effect in one example.

B-2.2 Coleman and Steele Approach

Coleman and Steele [4], expanding on the ideas advanced by Kline and McClintock [8] and assuming Gaussian error distributions, proposed viewing Eq. (B.14) as a propagation equation for 68% confidence intervals. A 95% coverage estimate of the uncertainty in the result was then proposed as that given by an equation similar to Eq. (B.23),

$$U_r^2 = B_r^2 + P_r^2 \tag{B.28}$$

with the systematic uncertainty of the result defined by

$$B_r^2 = \sum_{i=1}^{J} \theta_i^2 B_i^2 + 2 \sum_{i=1}^{J-1} \sum_{k=i+1}^{J} \theta_i \theta_k \rho_{B_{ik}} B_i B_k \tag{B.29}$$

and the random uncertainty (precision limit or precision uncertainty) of the result given by

$$P_r^2 = \sum_{i=1}^{J} \theta_i^2 (P_i)^2 + 2 \sum_{i=1}^{J-1} \sum_{k=i+1}^{J} \theta_i \theta_k \rho_{S_{ik}} P_i P_k \tag{B.30}$$

where $\rho_{B_{ik}}$ is the correlation coefficient appropriate for the systematic errors in X_i and in X_k and $\rho_{S_{ik}}$ is the correlation coefficient appropriate for the random errors in X_i and in X_k. The random uncertainty of the variable X_i is given by

$$P_i = t_i s_i \tag{B.31}$$

where t_i is determined with $\nu_i = N_i - 1$ degrees of freedom.

Equations (B.29) and (B.30) were viewed as propagation equations for 95% confidence systematic uncertainties and random uncertainties, and thus this approach avoided use of the Welch–Satterthwaite formula. Comparison of uncertainty coverages for a range of sample sizes using this approach and the Abernethy et al. approach have been presented [9]. As stated earlier in reference to the Abernethy et al. approach, consideration of these expressions in the context of the derivation presented in the preceding section shows that they cannot be justified on a rigorous basis.

Both the approach of Coleman and Steele [4] and the Abernethy et al. U_{RSS} approach [5] (properly modified to account for correlated systematic uncertainty effects) agree with the 95% confidence form of Eq. (B.18) for "large sample sizes"—that is, N_i (and ν_r) large enough so that t can be taken as 2.0.

B-2.3 ISO Guide Approach

The ISO guide [3] was published in late 1993 in the name of seven international organizations: the Bureau International des Poids et Mesures (BIPM), the International Electrotechnical Commission (IEC), the International Federation of Clinical Chemistry (IFCC), the International Organization for Standardization (ISO), the International Union of Pure and Applied Chemistry (IUPAC), the International Union of Pure and Applied Physics (IUPAP), and the International Organization of Legal Metrology (OIML). It is now the de facto international standard.

One fundamental difference between the approach of the guide and that of Eqs. (B.16), (B.18), and (B.19) is that the guide uses $u(x)$, a "standard uncertainty," to represent the quantities b_i and s_i used in this book. Instead of categorizing uncertainties as either systematic (bias) or random (precision), the u values are divided into type A standard uncertainties and type B standard uncertainties. *Type A uncertainties* are those evaluated "by the statistical analysis of series of observations," while *type B uncertainties* are those evaluated "by means other than the statistical analysis of series of observations." These do not correspond to the categories described by the traditional engineering usage: random (or precision or repeatability) uncertainty and systematic (or bias or fixed) uncertainty.

Arguments can, of course, be made for both sets of nomenclature. Type A and type B unambiguously define how an uncertainty estimate was made, whereas systematic and random uncertainties can change from one category to the other in a given experimental program depending on the experimental process used—a systematic calibration uncertainty can become a source of scatter (and thus a random uncertainty) if a new calibration is done before each reading in the sample is taken, for example. On the other hand, if one wants an estimate of the expected dispersion of results for a particular experimental approach or process, the systematic/random categorization is useful, particularly when used in the "debugging" phase of an experiment [10].

Considering the tradition in engineering of the systematic/random (bias/precision) uncertainty categorization and its usefulness in engineering experimentation as mentioned above, we have chosen to retain that categorization while adopting the mathematical procedures of the ISO guide. This categorization is also used in an AIAA standard [11], in AGARD [12] recommendations, and in the subsequent revisions [13] of Ref. 6.

B-2.4 AIAA Standard [11], AGARD [12], and ANSI/ASME [13] Approach

One of the authors (H.W.C.) was a participant in the North Atlantic Treaty Organization (NATO) Advisory Group for Aerospace Research and Development (AGARD) Fluid Dynamics Panel Working Group 15 on Quality Assessment for Wind Tunnel Testing and was the principal author of the methodology chapter in the resulting report [12]. This AGARD report, with minor revisions, was issued in 1995 as an AIAA standard [11]. The recommended methodology is that discussed in Section B-2.2, including the additional large sample assumptions discussed in Section B-3, so that t is taken as 2 "unless there are other overriding considerations" [12]. The other author (W.G.S.) is vice-chair of the ASME Committee PTC 19.1, which is responsible for the ANSI/ASME standard on test uncertainty. He was principal author of the revised methodology in the new standard. This revision also takes t as 2 for most engineering applications.

B-2.5 NIST Approach

Taylor and Kuyatt [14] reported guidelines for the implementation of a National Institute of Standards and Technology (NIST) policy, which states:

> Use expanded uncertainty U to report the results of all NIST measurements other than those for which u_c has traditionally been employed. To be consistent with current international practice, the value of k to be used at NIST for calculating U is, by convention, $k = 2$. Values of k other than 2 are only to be used for specific applications dictated by established and documented requirements.

(The coverage factor k corresponds to the K used here.) The NIST approach is thus that in the ISO guide [3], and no confidence level is associated with U when reported by NIST even though the coverage factor is specified as 2.0.

B-3 ADDITIONAL ASSUMPTIONS FOR ENGINEERING APPLICATIONS

In much engineering testing (e.g., as in most wind tunnel tests) it seems that the use of the complex but still approximate equations (B.18) and (B.19) for U_r and ν_r derived in Section B-1 would be excessively and unnecessarily complicated *and would tend to give a false sense of the degree of significance of the numbers computed using them* [11, 12]. In this section we examine what additional simplifying approximations can reasonably be made for application of uncertainty analysis in most engineering testing.

B-3.1 Approximating the Coverage Factor

Consider the process of estimating the uncertainty components b_i and s_i and obtaining U_r. The propagation equation and the Welch–Satterthwaite formula are approximate, not exact, and the Welch–Satterthwaite formula does not include the influence of correlated uncertainties. In addition, unavoidable uncertainties are always present in estimating the systematic standard uncertainties b_i and their associated degrees of freedom, v_{b_i}.

In fact, the uncertainty associated with an s_i calculated from N readings of X_i can be surprisingly large [3]. As shown in Figure B.2, for samples from a Gaussian parent population with standard deviation σ, 95 out of 100 determinations of s_i will scatter within an interval of approximately $\pm 0.45\sigma$ if the s_i are determined from $N = 10$ readings and within an interval of approximately $\pm 0.25\sigma$ if the s_i are determined from $N = 30$ readings. (A sample with $N \geq 31$ has traditionally been considered a "large" sample [6].) This effect seems to have received little consideration in the engineering "measurement uncertainty" literature. As stated in the ISO guide [3, pp.48–49]:

> This ... shows that the standard deviation of a statistically estimated standard deviation is not negligible for practical values of n. One may therefore conclude that type A evaluations of standard uncertainty are not necessarily more reliable than type B evaluations, and that in many practical measurement situations where the number of observations is limited, the components obtained from type B evaluations may be better known than the components obtained from type A evaluations.

Considering the 95% confidence t table (Table A.2), one can observe that for $v_r \geq 9$ the values of t are within about 13% of the large-sample t value

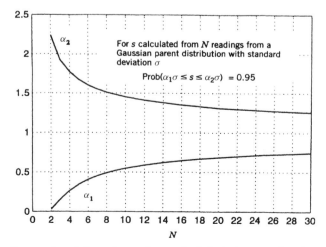

Figure B.2 Range of variation of sample standard deviation as a function of number of readings in the sample.

of 2. This difference is relatively insignificant compared with the uncertainties inherent in estimating the s_i and b_i as discussed above. *Therefore, for most engineering applications it is recommended that the central limit theorem be applied yielding a Gaussian error distribution for the result and an assumed value of* $t = 2$ *for a 95% level of confidence.* (This will be called the *large-sample assumption*.) This assumption eliminates the need for evaluation of ν_r using the Welch–Satterthwaite formula and thus the need to estimate all of the ν_{s_i} and the ν_{b_i}.

Consideration of the Welch–Satterthwaite formula [Eq. (B.19)] shows that because of the exponent of 4 in each term, ν_r is most influenced by the number of degrees of freedom of the largest of the $\theta_i s_i$ or $\theta_i b_i$ terms. If, for example, $\theta_3 s_3$ is dominant, then $\nu_r \approx \nu_{s_3} \geq 9$ for $N_3 \geq 10$ [recalling Eq. (B.20)]. If, on the other hand, $\theta_3 b_3$ is dominant, $\nu_r \approx \nu_{b_3} \geq 9$ when the relative uncertainty in b_3 is about 24% or less [recalling Eq. (B.21)]. If there is no single dominant term but there are M different $\theta_i s_i$ and $\theta_i b_i$ that all have the same magnitude and same number of degrees of freedom ν_a, then

$$\nu_r = M \nu_a \tag{B.32}$$

If $M = 3$, for example, ν_a would only have to be 3 or greater for ν_r to be equal to or greater than 9. Therefore, t can often legitimately be taken as 2 for estimating the uncertainty in a result determined from several measured variables even when the numbers of degrees of freedom associated with the measured variables are very small.

If the large-sample assumption is made so that $t = 2$, the 95% confidence expression for U_r becomes

$$U_r^2 = 2^2 \left[\sum_{i=1}^{J} \theta_i^2 b_i^2 + 2 \sum_{i=1}^{J-1} \sum_{k=i+1}^{J} \theta_i \theta_k b_{ik} \right.$$
$$\left. + \sum_{i=1}^{J} \theta_i^2 s_i^2 + 2 \sum_{i=1}^{J-1} \sum_{k=i+1}^{J} \theta_i \theta_k s_{ik} \right] \tag{B.33}$$

Thus for the large-sample case, we can define the systematic standard uncertainty of the result as

$$b_r^2 = \sum_{i=1}^{J} \theta_i^2 b_i^2 + 2 \sum_{i=1}^{J-1} \sum_{k=i+1}^{J} \theta_i \theta_k b_{ik} \tag{B.34}$$

and the random standard uncertainty of the result as

$$s_r^2 = \sum_{i=1}^{J} \theta_i^2 s_i^2 + 2 \sum_{i=1}^{J-1} \sum_{k=i+1}^{J} \theta_i \theta_k s_{ik} \tag{B.35}$$

and Eq. (B. 33) can be written as

$$U_r^2 = 2^2(b_r^2 + s_r^2) \tag{B.36}$$

with Eqs. (B.34) and (B.35) viewed as propagation equations for the systematic standard uncertainties and random standard uncertainties, respectively. The application of Eqs. (B.34) to (B.36) is termed *detailed uncertainty analysis* and is used once the planning phase of an experimental program is completed, as discussed in detail in Chapter 5.

If all the covariance terms in Eq. (B.33) are zero, the equation can be written as

$$U_r^2 = 2^2 \left[\sum_{i=1}^{J} \theta_i^2 b_i^2 + \sum_{i=1}^{J} \theta_i^2 s_i^2 \right] = \sum_{i=1}^{J} \theta_i^2 2^2 (b_i^2 + s_i^2) \tag{B.37}$$

or, most compactly,

$$U_r^2 = \sum_{i=1}^{J} \theta_i^2 U_i^2 \tag{B.38}$$

This equation describes the propagation of the overall uncertainties in the measured variables into the overall uncertainty of the result. Application of Eq. (B.38) is termed *general uncertainty analysis* and is used in the planning phase of an experiment, as described in detail in Chapter 4.

REFERENCES

1. Coleman, H. W., and Steele, W. G., "Engineering Application of Uncertainty Analysis," *AIAA Journal*, Vol. 33, No. 10, Oct. 1995, pp. 1888–1896.

2. Hildebrand, F. B., *Advanced Calculus for Applications*, Prentice-Hall, Upper Saddle River, NJ, 1962.

3. International Organization for Standardization (ISO), *Guide to the Expression of Uncertainty in Measurement*, ISO, Geneva, 1993. Corrected and reprinted, 1995.

4. Coleman, H. W., and Steele, W. G., "Some Considerations in the Propagation of Systematic and Random Errors into an Experimental Result," *Experimental Uncertainty in Fluid Measurements*, ASME FED Vol. 58, ASME, New York, 1987, pp. 57–62.

5. Abernethy, R. B., Benedict, R. P., and Dowdell, R. B., "ASME Measurement Uncertainty," *Journal of Fluids Engineering*, Vol. 107, 1985, pp. 161–164.

6. American National Standards Institute/American Society of Mechanical Engineers (ASME), *Measurement Uncertainty*, PTC 19.1-1985 Part 1, ASME, New York, 1986.

7. American National Standards Institute/American Society of Mechanical Engineers (ASME), *Measurement Uncertainty for Fluid Flow in Closed Conduits*, MFC-2M-1983, ASME, New York, 1984.

8. Kline, S. J., and McClintock, F. A., "Describing Uncertainties in Single-Sample Experiments," *Mechanical Engineering*, Vol. 75, 1953, pp. 3–8.

9. Steele, W. G., Taylor, R. P., Burrell, R. E., and Coleman, H. W., "Use of Previous Experience to Estimate Precision Uncertainty of Small Sample Experiments," *AIAA Journal*, Vol. 31, No. 10, 1993, pp. 1891–1896.

10. Coleman, H. W., Hosni, M. H., Taylor, R. P., and Brown, G. B., "Using Uncertainty Analysis in the Debugging and Qualification of a Turbulent Heat Transfer Test Facility," *Experimental Thermal and Fluid Science*, Vol. 4, No. 6, 1991, pp. 673–683.

11. American Institute of Aeronautics and Astronautics (AIAA), *Assessment of Wind Tunnel Data Uncertainty*, AIAA Standard S-071-1995, AIAA, New York, 1995.

12. Advisory Group for Aerospace Research and Development (AGARD), *Quality Assessment for Wind Tunnel Testing*, AGARD-AR-304, AGARD, Brussels, 1994.

13. American National Standards Institute/American Society of Mechanical Engineers (ASME), *Test Uncertainty*, PTC 19.1-2005, ASME, New York, 2005 (second revision of Ref. 6).

14. Taylor, B. N., and Kuyatt, C. E., "Guidelines for Evaluating and Expressing the Uncertainty of NIST Measurement Results," NIST Technical Note 1297, National Institute of Standards and Technology, Gaithersburg, MD, Sept. 1994, p. 5.

APPENDIX C

COMPARISON OF MODELS FOR CALCULATION OF UNCERTAINTY

In Appendix B we developed the equations that can be used to estimate the uncertainty of a determined result at any level of confidence. This methodology has been recommended by the International Organization for Standardization (ISO) in its guide [1] issued in 1993. Use of the ISO uncertainty methodology requires detailed knowledge of the variance and degrees of freedom for each uncertainty component; however, as shown in Appendix B, most engineering experiments can be analyzed satisfactorily by using the large-sample assumption for the uncertainty determination.

In this appendix we present the results of the Monte Carlo simulations that were performed to establish the validity of the large-sample model [2]. Six different data reduction equations were used in the simulations: mass balance, compressor efficiency, friction factor, mean of two temperatures, mean of five temperatures, and a single variable. For each equation the actual confidence levels attained were determined when the ISO models for 95% or 99% confidence were used and when the large-sample assumptions for these models were used.

C-1 MONTE CARLO SIMULATIONS

A Monte Carlo simulation technique was used to test the validity of the various uncertainty models for predicting the range around a result that contains the true result. For each example experiment, true values of the independent variables, X_i, were chosen and used to determine the true value of the experimental result. Also, the parent population distributions were chosen for the elemental systematic errors and the random errors. The random errors were always assumed to be from Gaussian distributions with assigned values of the parent population standard deviation, σ_{X_i}, for each variable. The elemental systematic error distributions were taken as either Gaussian or rectangular. In addition to the distribution type, the input quantity for the systematic errors was the plus or minus limit, B_{i_k}, on the parent distribution that contained 95% of the possible error values for each systematic error source. In appendix B (Section B-2) we defined this limit as the systematic uncertainty estimate using the nomenclature from the second edition of this book. For a Gaussian distribution, this 95% limit input quantity was equal to twice the standard deviation of the distribution. For a rectangular distribution, this 95% limit corresponded to 95% of the half-width of the distribution.

Each example experiment was simulated for N number of readings for each variable, where N was varied from 2 to 31. For a given case, N was the same for all variables; therefore, the random standard uncertainty for each variable had $N - 1$ degrees of freedom ν_{s_i}. Since the ISO models required a standard deviation and a number of degrees of freedom for each elemental systematic uncertainty, it was assumed that previous information was available to determine the systematic uncertainties. For each of the i variables, each of the k elemental systematic error distributions was sampled N times in order to determine a standard deviation with $N - 1$ degrees of freedom as

$$b_{i_k} = \left[\frac{1}{N-1} \sum_{n=1}^{N} [(\beta_{i_k})_n - \overline{\beta}_{i_k}]^2 \right]^{1/2} \tag{C.1}$$

The first reading from the N samples, $(\beta_{i_k})_1$, was used as the systematic error for that elemental source for that variable.

For each example and for each number of readings, N, 10,000 numerical experiments were run. First, for each experiment, the systematic errors, $(\beta_{i_k})_1$, and the standard deviations for the systematic errors, b_{i_k}, were determined. In these simulations, two elemental systematic errors were used for each variable. The parent population mean μ_{X_i} for each variable was then determined as

$$\mu_{X_i} = X_{i_{\text{true}}} + (\beta_{i_1})_1 + (\beta_{i_2})_1 \tag{C.2}$$

Using the value from Eq. (C.2) as the distribution mean and σ_{X_i} as the standard deviation, the Gaussian distribution for each variable was then sampled N times to get N values for each X_i. The mean, $\overline{X}$, and the sample standard deviation of the mean, $s_{\overline{X}_i}$, were then calculated for each variable. The result, r, was determined using these mean values. At this point the uncertainty interval from each model was determined, and a check was made for each model to see if r_{true} fell in the interval of r plus or minus the uncertainty interval. If it did, a counter was indexed for that model. The process was repeated 10,000 times for each value of N for each example experiment, allowing a coverage fraction to be determined for each uncertainty model.

The following expressions were used for the uncertainty models. The systematic uncertainty at the standard deviation level for each variable was determined as

$$b_i = [(b_{i_1})^2 + (b_{i_2})^2]^{1/2} \tag{C.3}$$

The b_{i_k} values and their degrees of freedom ($N - 1$ for these simulations) along with the corresponding random uncertainty information were then used directly in the Welch–Satterthwaite equation to determine the degrees of freedom in the result, ν_{ISO} [Eq. (B.19)], for the ISO models as

$$\nu_{\text{ISO}} = \frac{\left[\sum_{i=1}^{J} [(\theta_i b_i)^2 + (\theta_i s_{\overline{X}_i})^2] \right]^2}{\sum_{i=1}^{J} \left[(\theta_i s_{\overline{X}_i})^4 / \nu_{s_{\overline{X}_i}} + \sum_{k=1}^{2} [\theta_i b_{i_k}]^4 / (\nu_{b_{i_k}}) \right]} \tag{C.4}$$

where Eq. (B.19) has been expanded to include the elemental systematic standard uncertainties and their associated degree of freedom. The ISO uncertainties were then calculated as

$$U_{\text{ISO95}} = t_{\nu_{\text{ISO95}}} \left[\sum_{i=1}^{J} [(\theta_i s_{\overline{X}_i})^2 + (\theta_i b_i)^2] \right]^{1/2} \tag{C.5}$$

$$U_{\text{ISO99}} = t_{\nu_{\text{ISO99}}} \left[\sum_{i=1}^{J} [(\theta_i s_{\overline{X}_i})^2 + (\theta_i b_i)^2] \right]^{1/2} \tag{C.6}$$

For the constant models $t_{\nu_{\text{ISO95}}}$ was taken as 2.0 and $t_{\nu_{\text{ISO99}}}$ was taken as 2.6.

The discussion up to this point has considered the case where current information was known for the random uncertainty and previous information was known for the systematic error sources. Often, in practice, adequate current information is not available and previous information is needed for all the error sources [3]. This situation is definitely the case in single-reading tests; therefore, the study was extended to include the case where previous information was used to determine the random uncertainties.

Additional simulations were run where N previous samples were taken to calculate the sample standard deviation for each variable as

$$s_{X_i} = \left\{ \left[\frac{1}{N-1} \right] \sum_{k=1}^{N} (X_{i_k} - \overline{X}_i)^2 \right\}^{1/2} \tag{C.7}$$

Then one additional sample was taken to get the reading for X_i. The standard deviation s_{X_i} from Eq. (C.7) was used in all of the uncertainty and degrees of freedom calculations instead of the standard deviation of the mean. The main difference between the previous and current information cases was the larger random uncertainty in the previous information case because single readings instead of mean values were used for each parameter, and the $1/\sqrt{N}$ factor was not used in the standard deviation calculations.

For the current information cases for each example experiment, the relative contributions of the random and systematic errors were varied to see what effect they had on the confidence level for the uncertainty models. In the base or "balanced" case, the parent population standard deviations σ_{X_i} and the 95% systematic error estimates for the parent populations were chosen so that the relative contributions from the random and systematic error sources to the overall uncertainty were essentially the same for the case of 10 samples ($N = 10$).

One variation on the base case was to let the systematic uncertainty contribution dominate the uncertainty in the result. This domination was accomplished by multiplying the 95% systematic error estimates by a factor of 5 to 10 for each case. The magnitude of the factor was dependent on the data reduction equation for the example experiment. Another variation was to let the random uncertainty dominate the overall uncertainty. As in the systematic error–dominated case, the random error standard deviations were multiplied by a factor of 5 to 10. The last variation on the balanced case was to let the uncertainty in one variable dominate the overall uncertainty by setting all error estimates in the other variables equal to zero. This variation gave the smallest degrees of freedom for ν_{ISO} for each of the cases investigated.

For the previous information study, balanced (base), random error–dominated, and single variable–dominated cases were investigated. The parent population error estimates used in the balanced cases were the same as those used in the current information–balanced cases; therefore, these cases did not necessarily have equal random and systematic uncertainties at $N = 10$. The systematic error–dominated cases were not run, except for the friction factor example, since they would show results similar to those in the current information study. Also, since the previous information study affected only the treatment of the random uncertainty, the systematic error distributions were assumed Gaussian.

C-2 SIMULATION RESULTS

The data reduction equations for the six example experiments used in this study are given in Table C.1 along with true values of the variables and the parent population error estimates for each variable. These six equations provided a good mix of linear and nonlinear examples.

The results of the Monte Carlo simulations are presented in two ways. First the results are given for the degrees of freedom of the result approximately equal to nine, $\nu_r \approx 9$. This degrees of freedom is the ν_{ISO} value calculated from Eq. (C.4). For the simulations for each case for each data reduction equation at each value of N, the average of the 10,000 ν_{ISO} values was calculated and saved along with the uncertainty results. The simulation results shown first are those that

Table C.1 Data Reduction Equations, Hypothetical True Values of Test Variables, and Balanced Case Error Estimates for the Example Experiments

MASS BALANCE: $z = m_4 - m_1 - m_2 - m_3 = 0$

$m_1 = 50.0$ kg/h	$B_{m_{1_1}} = 1.0$ kg/h	$B_{m_{1_2}} = 3.0$ kg/h	$\sigma_{m_1} = 1.8$ kg/h
$m_2 = 50.0$ kg/h	$B_{m_{2_1}} = 1.0$ kg/h	$B_{m_{2_2}} = 3.0$ kg/h	$\sigma_{m_2} = 1.8$ kg/h
$m_3 = 50.0$ kg/h	$B_{m_{3_1}} = 1.0$ kg/h	$B_{m_{3_2}} = 3.0$ kg/h	$\sigma_{m_3} = 1.8$ kg/h
$m_4 = 150.0$ kg/h	$B_{m_{4_1}} = 1.0$ kg/h	$B_{m_{4_2}} = 3.0$ kg/h	$\sigma_{m_4} = 1.8$ kg/h

COMPRESSOR EFFICIENCY: $\eta = \left[(P_2/P_1)^{(\gamma-1)/\gamma} - 1\right]\left[(T_2/T_1) - 1\right]^{-1} = 0.87$

$P_1 = 101$ kPa	$B_{P_{1_1}} = 1.7$ kPa	$B_{P_{1_2}} = 2.6$ kPa	$\sigma_{P_1} = 4.8$ kPa
$P_2 = 659$ kPa	$B_{P_{2_1}} = 17.2$ kPa	$B_{P_{2_2}} = 10.3$ kPa	$\sigma_{P_2} = 31.7$ kPa
$T_1 = 294$ K	$B_{T_{1_1}} = 2.2$ K	$B_{T_{1_2}} = 1.6$ K	$\sigma_{T_1} = 4.3$ K
$T_2 = 533$ K	$B_{T_{1_1}} = 3.8$ K	$B_{T_{2_2}} = 3.3$ K	$\sigma_{T_2} = 7.9$ K
$\gamma = 1.4$			

FRICTION FACTOR: $f = (\pi^2 D^5 \Delta P)/(32\rho Q^2 \Delta X) = 0.005$

$D = 0.05$ m	$B_{D_1} = 0.5(10^{-3})$ m	$B_{D_2} = 2.5(10^{-4})$ m	$\sigma_D = 0.9(10^{-3})$ m
$\Delta P = 80.06$ Pa	$B_{\Delta P_1} = 2.5$ Pa	$B_{\Delta P_2} = 3.75$ Pa	$\sigma_{\Delta P} = 7.0$ Pa
$\rho = 1000.0$ kg/m^3	$B_{\rho_1} = 25.0$ kg/m^3	$B_{\rho_2} = 50.0$ kg/m^3	$\sigma_\rho = 90.0$ kg/m^3
$Q = 2.778(10^{-3})$ m^3/s	$B_{Q_1} = 0.5(10^{-4})$ m^3/s	$B_{Q_2} = 0.6(10^{-4})$ m^3/s	$\sigma_Q = 1.25(10^{-4})$ m^3/s
$\Delta X = 0.20$ m	$B_{\Delta X_1} = 5.0(10^{-3})$ m	$B_{\Delta X_2} = 1.0(10^{-2})$ m	$\sigma_{\Delta X} = 1.85(10^{-2})$ m

MEAN OF TWO TEMPERATURES: $\overline{T} = \frac{1}{2}(T_1 + T_2) = 305$ K

$T_1 = 300$ K	$B_{T_{1_1}} = 5.0$ K	$B_{T_{1_2}} = 3.5$ K	$\sigma_{T_1} = 9.5$ K
$T_2 = 310$ K	$B_{T_{2_1}} = 5.0$ K	$B_{T_{2_2}} = 3.5$ K	$\sigma_{T_2} = 9.5$ K

MEAN OF FIVE TEMPERATURES: $\overline{T} = \frac{1}{5}(T_1 + T_2 + T_3 + T_4 + T_5) = 305$ K

$T_1 = 300$ K	$B_{T_{1_1}} = 5.0$ K	$B_{T_{1_2}} = 3.5$ K	$\sigma_{T_1} = 9.5$ K
$T_2 = 310$ K	$B_{T_{2_1}} = 5.0$ K	$B_{T_{2_2}} = 3.5$ K	$\sigma_{T_2} = 9.5$ K
$T_3 = 307$ K	$B_{T_{3_1}} = 5.0$ K	$B_{T_{3_2}} = 3.5$ K	$\sigma_{T_3} = 9.5$ K
$T_4 = 305$ K	$B_{T_{4_1}} = 5.0$ K	$B_{T_{4_2}} = 3.5$ K	$\sigma_{T_4} = 9.5$ K
$T_5 = 303$ K	$B_{T_{5_1}} = 5.0$ K	$B_{T_{5_2}} = 3.5$ K	$\sigma_{T_5} = 9.5$ K

SINGLE VARIABLE: $X = 150$

$X = 150$	$B_{X_1} = 5.0$	$B_{X_2} = 3.5$	$\sigma_X = 9.5$

corresponded to the average $\nu_{\text{ISO}}(\nu_r)$ closest to 9. These results are given for the current information study in Table C.2 and for the previous information study in Table C.3. For each simulation type for each data reduction equation, the average ν_r is given along with the corresponding value of ν_{s_i} and $\nu_{b_{i_k}}$ ($N-1$ for these simulations) and the confidence level values determined for U_{ISO99}, $U_{2.6}$, U_{ISO95}, and $U_{2.0}$.

The second way that the simulation results are presented is for $N = 10$ or ν_{s_i} and $\nu_{b_{i_k}} = 9$. These results for each simulation type for each data reduction equation are given in Tables C.4 and C.5. Shown are the confidence levels for each uncertainty model along with the corresponding average value of ν_r for $N = 10$.

A review of the results for $\nu_r \approx 9$ in Tables C.2 and C.3 shows that the ISO models yield uncertainties with confidence levels close to the anticipated values,

Table C.2　Confidence Level (%) Results for Current Information Study with $\nu_r \approx 9$

Equation	ν_r	$\nu_{s_i}, \nu_{b_{i_k}}$	U_{ISO99}	$U_{2.6}$	U_{ISO95}	$U_{2.0}$
Eq. 1: mass balance						
Gauss.-bal	10.3	2	99.8	98.8	97.0	94.9
rec.-bal	10.3	2	99.8	98.8	96.9	94.8
Gauss.- B × 10	11.0	3	99.8	99.1	97.4	95.1
rec.-B × 10	7.1	2	99.8	98.7	98.1	94.7
Gauss.-σ × 10	9.4	3	99.4	97.8	96.2	93.4
rec.-σ × 10	9.7	3	99.5	98.2	95.9	93.6
Gauss.-m_4	9.0	5	99.8	99.0	97.8	95.1
rec.-m_4	9.6	6	99.9	99.6	98.5	96.9
Eq. 2: compressor efficiency						
Gauss.-bal	9.0	2	99.4	98.0	96.2	93.5
rec.-bal	10.5	2	99.4	98.4	96.4	94.1
Gauss.-B × 10	9.3	2	99.5	98.1	96.7	94.5
rec.-B × 10	9.8	2	99.4	98.0	96.8	94.6
Gauss.-σ × 6	8.8	3	99.1	97.6	96.3	93.7
rec.-σ × 6	8.9	3	99.3	97.4	96.0	93.5
Gauss.-P_2	10.2	5	99.2	97.7	95.4	93.0
rec.-P_2	8.6	4	99.5	97.8	95.8	93.6
Eq. 3: friction factor						
Gauss.-bal	10.4	2	99.2	97.9	95.8	93.7
rec.-bal	12.1	2	99.1	97.9	95.9	94.0
Gauss.-B × 5	12.3	2	98.5	97.0	95.3	93.9
rec.-B × 5	12.4	2	97.9	96.4	94.4	93.0
Gauss.-σ × 5	10.4	3	97.0	95.0	93.3	91.8
rec.-σ × 5	10.6	3	97.1	95.1	93.4	91.7
Gauss.-Q	10.6	5	99.1	97.8	95.5	93.4
rec.-Q	9.0	4	99.5	98.1	96.4	93.7
Eq. 4: mean of two temperatures						
Gauss.-bal	9.3	3	99.2	97.8	95.8	93.5
rec.-bal	10.8	3	99.4	98.2	95.8	93.6
Gauss.-B × 10	8.8	3	99.9	99.1	97.9	95.0
rec.-B × 10	9.2	3	99.9	99.1	98.0	95.4
Gauss.-σ × 10	8.9	5	99.1	97.3	95.2	92.3
rec.-σ × 10	9.0	5	99.1	97.3	95.5	92.8
Gauss.-T_1	8.0	4	99.3	97.5	95.6	92.2
rec.-T_1	8.9	4	99.5	98.1	96.2	93.6
Eq. 5: mean of five temperatures						
Gauss.-bal	10.8	2	99.5	98.3	96.0	93.9
rec.-bal	12.6	2	99.4	98.5	96.0	93.8
Gauss.-B × 10	11.6	2	99.8	99.0	97.0	95.0
rec.-B × 10	12.6	2	99.8	99.0	97.0	95.0
Gauss.-σ × 10	10.7	3	99.2	97.8	96.0	93.6
rec.-σ × 10	10.8	3	99.3	98.0	95.9	93.7
Gauss.-T_1	10.4	5	99.2	97.8	95.4	93.2
rec.-T_1	8.9	4	99.4	98.0	96.1	93.5
Eq. 6: single variable						
Gauss.-bal	7.9	4	99.3	97.4	95.7	92.9
rec.-bal	8.9	4	99.6	98.0	96.4	93.5
Gauss.-B × 7	9.0	5	99.9	99.3	97.9	95.3
rec.-B × 7	9.0	5	99.9	99.6	98.8	96.8
Gauss.-σ × 7	9.5	9	99.1	97.3	95.1	92.8
rec.-σ × 7	9.7	9	98.8	97.1	95.2	92.7

Table C.3 Confidence Level (%) Results for Previous Information Study with $\nu_r \approx 9$

Equation	ν_r	$\nu_{s_i}, \nu_{b_{i_k}}$	U_{ISO99}	$U_{2.6}$	U_{ISO95}	$U_{2.0}$
Eq. 1: mass balance						
Gauss.-bal	10.2	2	99.5	98.4	96.4	94.0
Gauss.-$\sigma \times 10$	8.9	3	99.3	97.7	96.3	93.2
Gauss.-m_4	9.2	9	99.0	97.2	94.9	92.4
Eq. 2: compressor efficiency						
Gauss.-bal	10.8	3	99.1	97.9	95.6	93.6
Gauss.-$\sigma \times 3$	8.8	3	98.9	97.1	95.9	93.4
Gauss.-P_2	8.9	7	98.8	96.8	94.5	91.7
Eq. 3: friction factor						
Gauss.-bal	7.8	2	99.2	97.1	95.8	93.0
Gauss.-$B \times 5$	14.2	2	97.9	96.6	94.6	93.4
Gauss.-Q	9.0	7	98.6	96.7	94.5	91.8
Eq. 4: mean of two temperatures						
Gauss.-bal	8.7	4	99.0	97.3	95.6	92.7
Gauss.-$\sigma \times 10$	8.8	5	99.3	97.6	95.8	93.1
Gauss.-T_1	9.0	7	98.8	97.3	95.1	92.6
Eq. 5: mean of five temperatures						
Gauss.-bal	8.0	2	99.6	98.0	96.6	93.5
Gauss.-$\sigma \times 10$	10.7	3	99.4	98.1	96.0	93.7
Gauss.-T_1	10.2	8	99.0	97.4	94.8	92.6
Eq. 6: single variable						
Gauss.-bal	9.0	7	98.8	97.0	95.0	92.1
Gauss.-$\sigma \times 3$	9.3	9	98.9	96.9	95.0	92.4
Gauss.-$B \times 4$	9.3	4	99.6	98.5	97.0	94.6

99% or 95%, as expected. However, the constant models also give uncertainties with confidence levels close to 99% or 95% even for the very small sample sizes ($N - 1$ or ν_{s_i} and $\nu_{b_{i_k}}$). As anticipated, the random uncertainty–dominated and one variable–dominated cases required the largest sample sizes to achieve $\nu_r \approx 9$.

The confidence level data given in Tables C.2 and C.3 are presented in histogram form in Figures C.1 and C.2 for the 99% study and the 95% study, respectively. It is seen in these figures that when $\nu_r \approx 9$ the 99% models (U_{ISO99} and $U_{2.6}$) yield confidence levels in the upper 90s and the 95% models (U_{ISO95} and $U_{2.0}$) yield confidence levels in the mid-90s. Considering the simplification in the uncertainty calculation resulting from using the constant models, it seems reasonable that if $\nu_r \approx 9$, then $U_{2.6}$ can be used for a 99% confidence uncertainty estimate and $U_{2.0}$ can be used for a 95% confidence uncertainty estimate.

The additional simulation results given in Tables C.4 and C.5 show that when the degrees of freedom are at least 9 for all the elemental uncertainty components ($\nu_{s_i} = \nu_{b_{i_k}} = 9$ or $N = 10$), the confidence levels obtained for both the

Table C.4 Confidence Level (%) Results for Current Information Study with Each ν_{s_i} and $\nu_{b_{i_k}} = 9$

Equation	ν_r	U_{ISO99}	$U_{2.6}$	U_{ISO95}	$U_{2.0}$
Eq. 1: mass balance					
Gauss.-bal	48.7	99.2	99.2	95.5	95.5
rec.-bal	48.1	99.3	99.3	95.4	95.4
Gauss.-B $\times$ 10	38.8	99.1	99.1	95.4	95.4
rec.-B $\times$ 10	41.1	99.4	99.4	95.6	95.6
Gauss.-σ $\times$ 10	36.5	98.7	98.7	94.7	94.7
rec.-σ $\times$ 10	39.0	99.0	99.0	95.2	95.1
Gauss.-m_4	14.6	99.9	99.4	97.0	95.6
rec.-m_4	13.3	100.0	99.8	98.6	97.4
Eq. 2: compressor efficiency					
Gauss.-bal	79.1	99.0	99.0	95.4	95.1
rec.-bal	88.3	99.1	99.1	95.7	95.7
Gauss.-B $\times$ 10	54.4	98.7	98.7	95.8	95.8
rec.-B $\times$ 10	56.4	98.4	98.4	95.3	95.3
Gauss.-σ $\times$ 6	32.9	98.5	98.4	94.5	94.4
rec.-σ $\times$ 6	33.8	98.7	98.6	95.2	95.1
Gauss.-P_2	21.0	99.2	98.7	95.6	94.8
rec.-P_2	22.2	99.6	99.2	96.2	95.5
Eq. 3: friction factor					
Gauss.-bal	92.8	99.0	99.0	95.3	95.3
rec.-bal	102.7	98.7	98.7	95.1	95.1
Gauss.-B $\times$ 5	64.8	97.2	97.2	94.2	94.2
rec.-B $\times$ 5	66.2	96.6	96.6	93.3	93.3
Gauss.-σ $\times$ 5	40.6	97.0	97.0	94.0	94.0
rec.-σ $\times$ 5	42.1	97.1	97.0	94.1	94.1
Gauss.-Q	22.2	99.1	98.7	95.3	94.3
rec.-Q	24.1	99.5	99.1	95.9	95.3
Eq. 4: mean of two temperatures					
Gauss.-bal	41.1	98.9	98.9	95.5	95.4
rec.-bal	45.1	99.2	99.2	95.7	95.7
Gauss.-B $\times$ 10	29.7	99.4	99.3	96.1	95.8
rec.-B $\times$ 10	30.9	99.7	99.6	95.9	95.7
Gauss.-σ $\times$ 10	16.9	99.0	98.1	95.2	94.0
rec.-σ $\times$ 10	17.1	99.2	98.3	95.1	94.0
Gauss.-T_1	21.7	99.2	98.7	95.5	94.8
rec.-T_1	23.3	99.6	99.1	95.8	95.1
Eq. 5: mean of five temperatures					
Gauss.-bal	98.4	99.1	99.1	95.2	95.2
rec.-bal	110.1	99.1	99.1	95.3	95.3
Gauss.-B $\times$ 10	70.2	99.1	99.1	95.4	95.4
rec.-B $\times$ 10	75.3	99.3	99.3	95.5	95.5
Gauss.-σ $\times$ 10	39.7	98.8	98.8	94.5	94.5
rec.-σ $\times$ 10	40.1	98.8	98.8	95.0	95.0
Gauss.-T_1	21.6	99.4	98.8	95.3	94.4
rec.-T_1	23.3	99.7	99.4	96.4	95.6
Eq. 6: single variable					
Gauss.-bal	21.7	99.3	98.8	95.6	94.8
rec.-bal	23.3	99.5	99.1	96.1	95.4
Gauss.-B $\times$ 7	16.3	99.8	99.4	96.7	95.3
rec.-B $\times$ 7	16.3	100.0	99.9	98.2	97.0
Gauss.-σ $\times$ 7	9.5	99.1	97.3	95.1	92.8
rec.-σ $\times$ 7	9.7	98.8	97.1	95.2	92.7

Table C.5 Confidence Level (%) Results for Previous Information Study with Each ν_{s_i} and $\nu_{b_{i_k}} = 9$

Equation	ν_r	U_{ISO99}	$U_{2.6}$	U_{ISO95}	$U_{2.0}$
Eq. 1: mass balance					
Gauss.-bal	64.9	99.0	99.0	95.0	95.0
Gauss.-$\sigma \times 10$	31.9	98.9	98.8	94.8	94.6
Gauss.-m_4	9.2	99.0	97.2	94.9	92.4
Eq. 2: compressor efficiency					
Gauss.-bal	38.2	98.8	98.8	94.7	94.7
Gauss.-$\sigma \times 3$	31.2	98.4	98.1	95.0	94.8
Gauss.-P_2	11.3	99.1	97.8	95.2	93.0
Eq. 3: friction factor					
Gauss.-bal	46.1	98.1	98.1	94.6	94.6
Gauss.-$B \times 5$	90.6	96.7	96.7	93.5	93.5
Gauss.-Q	11.4	98.8	97.6	94.9	93.0
Eq. 4: mean of two temperatures					
Gauss.-bal	20.5	99.0	98.2	95.0	93.9
Gauss.-$\sigma \times 10$	16.6	99.0	98.1	95.2	93.7
Gauss.-T_1	11.4	98.8	97.4	94.8	93.1
Eq. 5: mean of five temperatures					
Gauss.-bal	47.3	98.7	98.7	94.9	94.9
Gauss.-$\sigma \times 10$	38.9	98.8	98.8	94.8	94.8
Gauss.-T_1	11.4	98.9	97.6	95.2	93.1
Eq. 6: single variable					
Gauss.-bal	11.4	98.9	97.5	94.9	92.9
Gauss.-$\sigma \times 3$	9.3	98.9	96.9	95.0	92.4
Gauss.-$B \times 4$	22.8	99.4	98.8	95.7	95.1

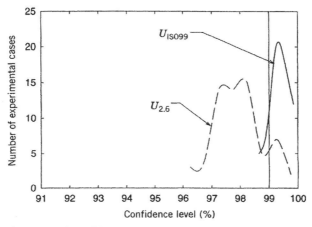

Figure C.1 Histogram of confidence levels provided by 99% uncertainty models for all example experiments when $\nu_r \approx 9$.

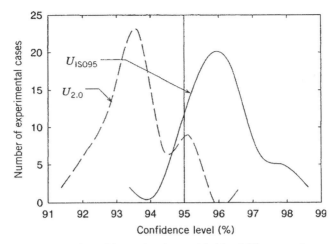

Figure C.2   Histogram of confidence levels provided by 95% uncertainty models for all example experiments when $\nu_r \approx 9$.

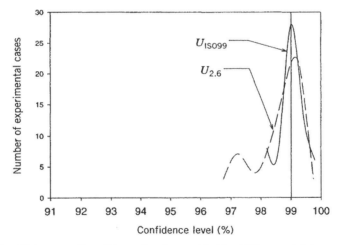

Figure C.3   Histogram of confidence levels provided by 99% uncertainty models for all example experiments when $\nu_{s_i} = \nu_{b_{i_k}} = 9$.

ISO models and the constant models are even closer to the anticipated values. The histograms for these results in Figures C.3 and C.4 show the 95% models approximately centered around 95 and the 99% models around 99. Of course, the degrees of freedom of the result, ν_r, is greater than 9 for all these cases, as shown in Tables C.4 and C.5. These simulation results support those presented above for $\nu_r \approx 9$ and show that the large-sample assumption applies for most

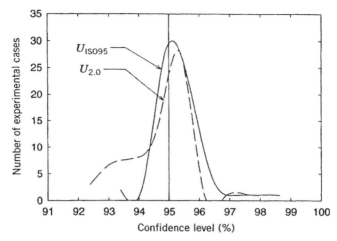

Figure C.4 Histogram of confidence levels provided by 95% uncertainty models for all example experiments when $\nu_{s_i} = \nu_{b_{i_k}} = 9$.

engineering and scientific applications, and the constant models for uncertainty determination are definitely appropriate.

REFERENCES

1. International Organization for Standardization (ISO), *Guide to the Expression of Uncertainty in Measurement*, ISO, Geneva, 1993.
2. Steele, W. G., Ferguson, R. A., Taylor, R. P., and Coleman, H. W., "Comparison of ANSI/ASME and ISO Models for Calculation of Uncertainty," *ISA Transactions*, Vol. 33, 1994, pp. 339–352.
3. Steele, W. G., Taylor, R. P., Burrell, R. E., and Coleman, H. W., "Use of Previous Experience to Estimate Precision Uncertainty of Small Sample Experiments," *AIAA Journal*, Vol. 31, No. 10, Oct. 1993, pp. 1891–1896.

APPENDIX D

SHORTEST COVERAGE INTERVAL FOR MONTE CARLO METHOD

In Section 3-2.3, the methodology is given for determining coverage intervals for Monte Carlo Method (MCM) simulations when the distribution of the results is asymmetric. The approach shown is the probabilistically symmetric coverage interval. The GUM supplement [1] provides an alternative coverage interval for MCM simulations called the "shortest coverage interval." This interval is the shortest among all possible intervals for a distribution of MCM results where each interval has the same coverage probability. The concept is illustrated in Figure D.1.

For a distribution of M Monte Carlo results, there can be $(1 - p)M$ intervals that contain $100p\%$ of the total number of results. In Figure D.1, we show the first interval, which starts with sorted result 1, and the last interval, which ends with sorted result M. For instance, if we wanted a 95% level of coverage,

$$\text{Interval } 1 = \text{result}(1 + pM) - \text{result } 1 \tag{D.1}$$

where $p = 0.95$ for 95% coverage and the last interval would be

$$\text{Interval } (1 - p)M = \text{result } (M) - \text{result } (1 - p)M \tag{D.2}$$

The other intervals would be

$$\text{Interval } 2 = \text{result } [(1 + pM) + 1] - \text{result } 2 \tag{D.3}$$

$$\text{Interval } 3 = \text{result } [(1 + pM) + 2] - \text{result } 3 \tag{D.4}$$

and so on. The interval with the smallest value is then the shortest coverage interval with a $100p\%$ level of confidence. As discussed for the probabilistically

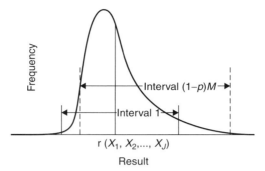

Figure D.1   Concept for determining shortest coverage interval for a $100p\%$ level of confidence.

symmetric coverage interval in Section 3-2.3, the appropriate integer values are used for the result numbers.

The GUM supplement [1] states that the shortest coverage interval will contain the mode, or the most probable value of the distribution of results. If the distribution of MCM results is symmetric, the shortest coverage interval, the probabilistically symmetric coverage interval, and the interval given by $\pm(k_{100p\%})s_{MCM}$ will be the same for a $100p\%$ level of confidence. The appropriate model to use to determine the uncertainty is up to the user.

REFERENCE

1. Joint Committee for Guides in Metrology (JCGM), "Evaluation of Measurement Data—Supplement 1 to the 'Guide to the Expression of Uncertainty in Measurement'—Propagation of Distributions Using a Monte Carlo Method," JCGM 101:2008, France, 2008.

APPENDIX E

ASYMMETRIC SYSTEMATIC UNCERTAINTIES

In some experimental situations, there are significant systematic errors that displace the experimental result in a specific direction away from the true result. These asymmetric systematic errors may vary as the experimental result varies, but they are fixed for a given value of the result. Moffat [1] calls this type of error "variable but deterministic."

Transducer installation systematic errors discussed in Chapter 5 are usually of this asymmetric type. They tend to cause the transducer output to be consistently either higher or lower than the quantity that is being measured. If the effects of these asymmetric systematic errors are significant, then they must be estimated and included in the determination of the systematic uncertainty of the result. Note that in some cases the effect of these asymmetric errors may not be large enough to justify the effort of dealing with them rigorously. In these cases, an effective symmetric uncertainty limit is used. The methodology for handling asymmetric systematic uncertainties is given below for TSM propagation in Section E-1 and for MCM propagation in Section E-2 and is then illustrated in the examples that follow.

In many cases an analytical expression can be formulated for these asymmetric systematic errors, and this expression can be included in the data reduction equation [as in the case of conduction and radiation losses in Section 5-6, Eq. (5.72)]. This technique will usually allow us to reduce the overall uncertainty in the experimental result. Instead of the original asymmetric systematic uncertainty, the symmetric systematic and random uncertainties in the new terms of the data reduction expression become factors in determining the uncertainty. This concept is illustrated in the following sections.

E-1 PROCEDURE FOR ASYMMETRIC SYSTEMATIC UNCERTAINTIES USING TSM PROPAGATION

Normally, the plus and minus limit estimates for a systematic uncertainty interval are equal, and for no random uncertainty, we assume that the true value lies somewhere in this symmetric distribution around the reading. There are situations, however, when we have sufficient information to know that the truth is more likely to be on one side of the reading than the other. An example is the radiation error that occurs when a thermocouple is used to measure the temperature of a hot gas flowing in an enclosure with cooler walls. The thermocouple is being heated by convection from the gas, but it is losing heat by radiation to the cooler walls. If the radiation error is a significant component of the systematic uncertainty, the true gas temperature is more likely to be above the thermocouple temperature reading, and the systematic uncertainty for this measurement would be larger in the positive direction than in the negative direction around the temperature reading.

In some experiments, physical models can be used to replace the asymmetric uncertainties with symmetric uncertainties in additional experimental variables. For instance, when performing an energy balance on the thermocouple probe just discussed, the convective heat transfer to the probe is equal to the radiation heat loss to the walls at steady-state conditions, assuming negligible conduction loss from the probe:

$$h(T_g - T_t) = \epsilon\sigma(T_t^4 - T_w^4) \tag{E.1}$$

where T_t, T_g, and T_w are the thermocouple probe, gas, and wall temperatures, respectively; h is the convective heat transfer coefficient; ϵ is the emissivity of the probe surface; and σ is the Stefan–Boltzmann constant. Solving this equation for T_g yields the new data reduction equation:

$$T_g = T_t + \frac{\epsilon\sigma}{h}(T_t^4 - T_w^4) \tag{E.2}$$

The asymmetric uncertainty in using T_t as the estimate of the gas temperature has been replaced by additional experimental variables that have symmetric uncertainties. This technique might reduce the overall uncertainty in T_g, depending on the uncertainties in ϵ, h, and T_w.

When possible, the "zero centering" of the uncertainty using appropriate physical models as shown above should be used when asymmetric systematic uncertainties are present. There may be cases, however, where the uncertainties introduced by the additional variables in the data reduction equation yield a larger uncertainty interval than the original asymmetric uncertainty interval or where the experimenter decides that asymmetric uncertainties are more appropriate. In those cases, the methods presented in this appendix should be used [2].

Consider an experimental result that is a function of X and Y as

$$r = r(X, Y) \tag{E.3}$$

where for simplicity we will assume that X and Y have no random uncertainty. Let us also assume that the variable X has two symmetric systematic uncertainty components b_{X_1} and b_{X_2} and that variable Y has two symmetric components b_{Y_1} and b_{Y_2}. None of the error sources are common for X and Y, so there are no correlated systematic errors.

For X we consider an additional asymmetric systematic error β_3 with possible distributions as shown in Figures E.1 to E.3 depending on the assumed distribution of the error. For each of the distributions (Gaussian, rectangular, or triangular), we have a lower limit (LL) and an upper limit (UL) around X which define the extent of the distribution. For the Gaussian distribution, we take the interval $X - LL$ to $X + UL$ to contain 95% of the distribution. For the rectangular and triangular distributions, this interval defines the lower and upper extremes of the distribution. The triangular distribution has an additional limit, MPL, representing the interval between the mode, or most probable point, of the distribution and the measurement X. All of the limits, UL, LL, and MPL, are positive numbers. Note that for the Gaussian distribution (Figure E.1) the mode is fixed once X, LL, and UL are known. Therefore, the rectangular distribution (Figure E.2) may be more appropriate to use if there is no estimate of the mode. If there is a reasonable estimate of the mode, then the triangular distribution (Figure E.3) may be the best to use.

The procedure for handling the asymmetric uncertainty is the same for each of the distributions in Figures E.1 to E.3. First an interval is determined that is the difference, c_X, between the distribution mean and the measurement X. The

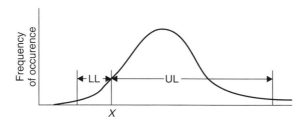

Figure E.1   Gaussian asymmetric systematic error distribution for β_3.

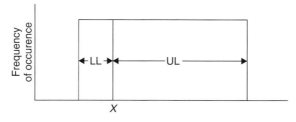

Figure E.2   Rectangular asymmetric systematic error distribution for β_3.

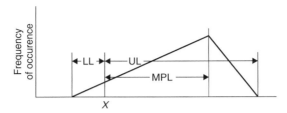

Figure E.3   Triangular asymmetric systematic error distribution for β_3.

Table E.1   Expressions for c_X for Gaussian, Rectangular, and Triangular Distributions in Figures E.1 through E.3

Distribution	c_X
Gaussian	$\frac{1}{2}(\text{UL} - \text{LL})$
Rectangular	$\frac{1}{2}(\text{UL} - \text{LL})$
Triangular	$\frac{1}{3}(\text{UL} - \text{LL} + \text{MPL})$

expressions for c_X for the three distributions are given in Table E.1. Note that c_X can be positive or negative depending on the relative values of UL, LL, and MPL. If X is greater than the mean, c_X will be negative, and if X is less than the mean, c_X will be positive.

The next step is to determine the standard deviation b_{X_3}, for the asymmetric error distribution. These expressions are given in Table E.2. This elemental systematic standard uncertainty is combined with the other elemental systematic uncertainties for X using Eq. (5.16) to obtain

$$b_X = \left[b_{X_1}^2 + b_{X_2}^2 + b_{X_3}^2 \right]^{1/2} \tag{E.4}$$

The similar expression for b_Y is

$$b_Y = \left[b_{Y_1}^2 + b_{Y_2}^2 \right]^{1/2} \tag{E.5}$$

Table E.2   Standard Deviation b_{X_3} for Gaussian, Rectangular, and Triangular Distributions in Figures E.1 through E.3

Distribution	b_{X_3}
Gaussian	$\frac{1}{4}(\text{UL} + \text{LL})$
Rectangular	$\dfrac{\text{UL} + \text{LL}}{2\sqrt{3}}$
Triangular	$\left[\dfrac{\text{UL}^2 + \text{LL}^2 + \text{MPL}^2 + (\text{LL})(\text{UL}) + (\text{LL})(\text{MPL}) - (\text{UL})(\text{MPL})}{18} \right]^{1/2}$

The systematic standard uncertainty of the result is then calculated using Eq. (3.11) as

$$b_r = [(\theta_X b_X)^2 + (\theta_Y b_Y)^2]^{1/2} \tag{E.6}$$

where the derivatives θ_X and θ_Y are evaluated at the point $((X + c_X), Y)$.

As shown in Ref. 2, the true value of r for this case with no random uncertainty will fall with a 95% level of confidence in the interval

$$r((X + c_X), Y) - 2b_r \leq r_{\text{true}} \leq r((X + c_X), Y) + 2b_r \tag{E.7}$$

Now if a factor F is defined so that

$$F = r((X + c_X), Y) - r(X, Y) \tag{E.8}$$

then the interval for r_{true} can be written as

$$r(X, Y) - (2b_r - F) \leq r_{\text{true}} \leq r(X, Y) + (2b_r + F) \tag{E.9}$$

The asymmetrical interval $-(2b_r - F)$ and $+(2b_r + F)$ is the systematic uncertainty interval for the result $r(X, Y)$.

At first thought it might seem more reasonable to add the factor c_X to variable X and to report the result as $r((X + c_X), Y)$ with a symmetric uncertainty interval; however, this technique is not recommended because c_X is based on an estimated systematic uncertainty and not on a known systematic calibration error. Therefore, it is more correct to report the result determined at the measured variables and the associated asymmetric uncertainty interval.

For the general case, where

$$r = r(X_1, X_2, \ldots, X_J) \tag{E.10}$$

the total uncertainty for the result is determined as

$$U_r = 2 \left[\sum_{i=1}^{J} (\theta_i b_i)^2 + 2 \sum_{i=1}^{J-1} \sum_{k=i+1}^{J} \theta_i \theta_k b_{ik} + \sum_{i=1}^{J} (\theta_i s_i)^2 \right]^{1/2} \tag{E.11}$$

where in each b_i where there is an elemental asymmetric systematic uncertainty, an appropriate standard deviation is used, as shown in Table E.2. The true result will be in the interval

$$r(X_1, X_2, \ldots, X_J) - (U_r - F) \leq r_{\text{true}} \leq r(X_1, X_2, \ldots, X_J) + (U_r + F) \tag{E.12}$$

where F is calculated as shown in Eq. (E.8) with a factor c_{X_1} used for each variable that has asymmetric systematic uncertainty. As shown in Eq. (E.12), the total uncertainty components for the result are

$$U_r^+ = U_r + F \tag{E.13}$$

$$U_r^- = U_r - F \tag{E.14}$$

E-2 PROCEDURE FOR ASYMMETRIC SYSTEMATIC UNCERTAINTIES USING MCM PROPAGATION

The distributions given in Figures E.1 through E.3 are essentially distributions for the modified variable $X + c_X$. The mean of each distribution is $x_{\text{true}} + c_X$, where here x_{true} is taken as the nominal value for X, as was done for MCM inputs in Chapter 3. Considering a specific error $\beta_3(i)$ from an iteration of the MCM propagation, the value of $(X + c_X)(i)$ would be

$$(X + c_X)(i) = X_{\text{true}} + c_X + \beta_1(i) + \beta_2(i) + \beta_3(i) \tag{E.15}$$

where $\beta_1(i)$ and $\beta_2(i)$ are the errors taken from the distributions representing the symmetric error sources 1 and 2 for X discussed in Section E-1.

The MCM then follows the same process as that shown in Figure 3.4 with the result determined as

$$r[(X + c_X), Y](i) = f[(X + c_X)(i), Y(i)] \tag{E.16}$$

The standard deviation of the converged set of results will be the b_r given in Eq. (E.6). The procedures in Eqs. (E.7) through (E.14) would then be used to determine the range for r_{true} and to evaluate the asymmetric uncertainty limits for the result.

Note that in the MCM flowchart given in Figure 3.4 the error distributions have mean values of zero. Therefore, care must be taken to properly use the random number generators for the asymmetric error distributions. For Gaussian and rectangular distributions, means of zero and the standard deviations from Table E.2 would be the inputs to the random number generator. For a triangular random number generator, the inputs would typically be the lower extreme a, the mode c, and the upper extreme b as defined below:

$$a = X_{\text{true}} - \text{LL} \tag{E.17}$$

$$c = X_{\text{true}} + \text{MPL} \tag{E.18}$$

$$b = X_{\text{true}} + \text{UL} \tag{E.19}$$

Samples from this distribution will be values $X_{\text{true}} + c_X + \beta_3(i)$ to which the errors $\beta_1(i)$ and $\beta_2(i)$ are added to get the iteration value of $(X + c_X)(i)$ as shown in Eq. (E.15).

In the following section we present an example that illustrates the methodology for dealing with asymmetric systematic uncertainties. The example uses the TSM approach.

E-3 EXAMPLE: BIASES IN A GAS TEMPERATURE MEASUREMENT SYSTEM

A thermocouple probe with a $\frac{1}{8}$ -in.-diameter stainless-steel sheath is inserted into the exhaust of a 50-hp diesel engine as shown in Figure E.4. The exhaust pipe is a 4-in.-diameter cylinder. We wish to determine the exhaust gas temperature.

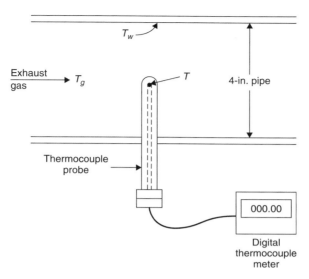

Figure E.4   Thermocouple probe in engine exhaust pipe.

Before conducting an analysis of the measurement system for determining the gas temperature T_g, we must first estimate the conditions in the diesel exhaust. A standard textbook on internal combustion engines [3] can provide the required information. For a diesel engine operating at full throttle and loaded to 50 hp, the fuel flow rate would be about 25 lbm/h and the air flow rate would be about 375 lbm/h. The exhaust temperature would be in the neighborhood of 1000°F (1460°R) at the exhaust manifold. Assuming that the T_g measurement is to be made in the pipe near the exhaust manifold, the gas temperature would be about 1000°F and the pipe wall temperature would be lower, perhaps 750°F.

The measurement system for this temperature determination is a standard stainless-steel-sheathed thermocouple connected to a digital thermometer. The thermocouple bead is located at the end of the probe in the center of the pipe. For this example we assume that this system has not been calibrated, but information is available about the accuracy of these devices. The manufacturer's accuracy specification for the thermocouple is ±2.2°C or ±0.75% of the temperature, whichever is greater [4]. The percentage in this case refers to the temperature in degrees Celsius rather than an absolute temperature as confirmed in a telephone conversation with the manufacturer. Converting this specification to degrees Fahrenheit gives an accuracy of ±4.0°F or ±0.75% of the temperature in °F − 32°F [±0.0075(T − 32°F)], whichever is greater. The manufacturer's accuracy specification for the digital thermometer is ±2°F.

We will take these specifications to represent 95% systematic uncertainties in the temperature determination. Therefore, for this example, the systematic standard uncertainty for the thermocouple probe would be (assuming a Gaussian distribution)

$$b_{\text{probe}} = \tfrac{1}{2}(0.0075)(1000°\text{F} - 32°\text{F}) = 3.6°\text{F} \qquad \text{(E.20)}$$

Combining the probe and thermometer uncertainties ($b_{thm} = 1°F$) gives us a systematic standard uncertainty estimate for the gas temperature measurement of

$$b_{T_g} = [(3.6)^2 + (1)^2]^{1/2} = 3.7°F \tag{E.21}$$

Neglecting random uncertainties, 2 times this value, or 7.4°F, might be an initial estimate of the range around the probe temperature T_t for a 95% coverage interval for the gas temperature. However, in this case there are other physical processes occurring that will significantly affect the measurement.

Other bias errors influence the temperature determination. There will be some variation in the gas temperature around the pipe center, but for this example we are considering the gas temperature around the tip of the thermocouple probe; therefore, we consider any temperature variations in this region to be negligible.

A significant bias error will occur in the temperature determination because of the radiation heat loss from the probe to the pipe wall. With the hot gas and the cooler pipe wall, the thermocouple will be at a temperature between these two limits. For this example, the thermocouple probe temperature T_t could be as much as, say, 60°F below the gas temperature. This estimate would be an asymmetric systematic uncertainty since T_t would always be lower than T_g by this amount for a given operating condition.

Taking all of these systematic uncertainty estimates into account, the systematic uncertainty for the gas temperature determination is evaluated by first defining the limits for the radiation error as

$$UL_{radiation} = 60°F \tag{E.22}$$

$$LL_{radiation} = 0°F \tag{E.23}$$

Assuming a rectangular distribution for the radiation error, the standard deviation from Table E.2,

$$\begin{aligned} b_{radiation} &= \frac{UL_{radiation} + LL_{radiation}}{2\sqrt{3}} \\ &= \frac{60°F + 0°F}{2\sqrt{3}} \\ &= 17.3°F \end{aligned} \tag{E.24}$$

This uncertainty is combined with the other elemental systematic uncertainty estimates for the gas temperature to determine b_{T_g} as

$$\begin{aligned} b_{T_g} &= (b_{probe}^2 + b_{meter}^2 + b_{radiation}^2)^{1/2} \\ &= [(3.7)^2 + (1)^2 + (17.3)^2]^{1/2} \\ &= 17.7°F \end{aligned} \tag{E.25}$$

The values for c_{T_g} and F are evaluated using Table E.1 and Eq. (E.8),

$$c_{T_g} = \tfrac{1}{2}(\text{UL}_{\text{radiation}} - \text{LL}_{\text{radiation}})$$
$$= \tfrac{1}{2}(60°\text{F} - 0°\text{F})$$
$$= 30°\text{F} \qquad (\text{E}.26)$$

and

$$F = T_t + 30°\text{F} - T_t = 30°\text{F} \qquad (\text{E}.27)$$

The uncertainty components for T_g are then calculated as (where we have assumed no random uncertainty)

$$U_{T_g}^+ = 2b_{T_g} + F = 35.4°\text{F} + 30°\text{F} = 65.4°\text{F} \qquad (\text{E}.28)$$
$$U_{T_g}^- = 2b_{T_g} - F = 35.4°\text{F} - 30°\text{F} = 5.4°\text{F} \qquad (\text{E}.29)$$

These asymmetric uncertainty limits can be used to find the total uncertainty interval with 95% level of confidence for the true gas temperature, so that

$$T_t - U_{T_g}^- \le T_g \le T_t + U_{T_g}^+$$
$$1000°\text{F} - 5.4°\text{F} \le T_g \le 1000°\text{F} + 65.4°\text{F} \qquad (\text{E}.30)$$
$$995°\text{F} \le T_g \le 1065°\text{F}$$

This interval is much larger than our initial estimate of the range for the gas temperature of $T_t \pm 7.4°\text{F}$. This example illustrates the consequences of neglecting a significant physical bias when making a measurement.

Depending on the application for this temperature determination, the uncertainty estimates above may or may not be acceptable. The significant systematic uncertainty in this case is, of course, the radiation effect. This example is a case in which the physical principles associated with the process, the radiation heat loss, can be incorporated into the data reduction equation to "zero center" the uncertainty and possibly reduce the overall uncertainty in the final determination.

In Section E-1 we considered an energy balance on the thermocouple probe where

$$\text{energy in} = \text{energy out} \qquad (\text{E}.31)$$

or

$$\text{convection to probe} = \text{radiation from probe}$$
$$+ \text{ conduction out of probe} \qquad (\text{E}.32)$$

The conduction term is much smaller than the radiation and will be neglected, and the resulting expression for T_g from Eq. (E.2) yields the data reduction equation

$$T_g = \frac{\epsilon \sigma}{h}(T_t^4 - T_w^4) + T_t$$

Note that absolute temperatures in Rankine are used in this expression.

Uncertainty analysis can now be used to estimate the uncertainty in the gas temperature determination from the uncertainties associated with T_t, T_w, ϵ, and h. We will still make the assumption that the systematic uncertainties are much greater than the random uncertainties, so that $2b_{T_g} = U_{T_g}$. The uncertainty expression for this example is

$$b_{T_g} = \left[\left(\frac{\partial T_g}{\partial \epsilon} b_\epsilon \right)^2 + \left(\frac{\partial T_g}{\partial h} b_h \right)^2 + \left(\frac{\partial T_g}{\partial T_t} b_{T_t} \right)^2 + \left(\frac{\partial T_g}{\partial T_w} b_{T_w} \right)^2 \right]^{1/2} \quad \text{(E.33)}$$

where

$$\frac{\partial T_g}{\partial \epsilon} = \frac{\sigma}{h}(T_t^4 - T_w^4) \quad \text{(E.34)}$$

$$\frac{\partial T_g}{\partial h} = \frac{-\epsilon \sigma}{h^2}(T_t^4 - T_w^4) \quad \text{(E.35)}$$

$$\frac{\partial T_g}{\partial T_t} = \frac{4\epsilon \sigma}{h} T_t^3 + 1 \quad \text{(E.36)}$$

$$\frac{\partial T_g}{\partial T_w} = \frac{-4\epsilon \sigma}{h} T_w^3 \quad \text{(E.37)}$$

Values for the emissivity and the convective heat transfer coefficient are required. These can be obtained from a standard heat transfer textbook [5]. Using a heat transfer correlation for crossflow over a cylinder, a value for h is determined to be 47 Btu/hr-ft^2-$^\circ$R. It is known that these heat transfer correlations are accurate to about ±25%; therefore, for h the 95% systematic uncertainty, $2b_h$, is about 12 Btu/hr-ft^2-$^\circ$R.

The emissivity of a metal surface is very dependent on the surface condition. For the stainless-steel (type 304) thermocouple probe, the emissivity can vary from 0.36 to 0.73 [6]. Therefore, for our initial analysis we will take ϵ as 0.55 with a 95% systematic uncertainty of 0.19.

The pipe wall temperature has been estimated to be about 750°F (1210°R), but this value could be in error by as much as ±100°R. This estimate gives us a value for $2b_{T_w}$. The systematic, standard uncertainty in the temperature measurement, T_t, is the combination of the probe and meter uncertainties, so that [Eq. (E. 21)]

$$b_{T_t} = [(3.6)^2 + (1)^2]^{1/2} = 3.7°F$$

In summary, the nominal values and standard uncertainty estimates (assuming Gaussian distributions) for the quantities used to determine the gas temperature are

$$\epsilon = 0.55 \pm 0.095 \qquad (E.38)$$

$$h = 47 \pm 6 \text{ Btu/hr-ft}^2\text{-}^\circ\text{R} \qquad (E.39)$$

$$T_t = 1460 \pm 3.7^\circ\text{R} \qquad (E.40)$$

$$T_w = 1210 \pm 50^\circ\text{R} \qquad (E.41)$$

Using the data reduction equation [Eq. (E.2)], the uncertainty expression [Eq. (E.33)], and $U_{T_g} = 2b_{T_g}$, the gas temperature would be

$$T_g = 1508 \pm 27^\circ\text{R} \qquad (E.42)$$

This value is an improvement over our original calculation, in which the radiation effect was estimated. But it still might not be good enough, depending on the application.

Let us examine the calculation of the systematic standard uncertainty in T_g. Substituting the nominal values for ϵ, h, T_t, and T_w along with their systematic standard uncertainties into the uncertainty expression yields

$$\begin{array}{cccc} (\epsilon) & (h) & (T_t) & (T_w) \\ b_{T_g} = [69 & + \quad 38 & + \quad 21 & + \quad 50]^{1/2} \end{array} \qquad (E.43)$$

The uncertainty in the emissivity is the largest contributor to the uncertainty in T_g. A closer examination of the available data for ϵ [6] shows that for type 304 stainless steel exposed to a temperature near 1000°F for over 40 h the emissivity varies from 0.62 to 0.73. Therefore, after this thermocouple probe has been in service for some time, the value of ϵ would be about 0.68 ± 0.028.

These revised values yield the following estimates:

$$\begin{array}{cccc} (\epsilon) & (h) & (T_t) & (T_w) \\ b_{T_g} = [7 & + \quad 58 & + \quad 25 & + \quad 77]^{1/2} \\ & & = 13^\circ\text{R} \end{array} \qquad (E.44)$$

and

$$T_g = 1520 \pm 26^\circ\text{R} \qquad (E.45)$$

The contribution of the systematic uncertainty in ϵ to the systematic uncertainty in the gas temperature has been reduced significantly, but the increase in the nominal value of ϵ has increased the other terms in the uncertainty expression. The overall uncertainty in the T_g determination is essentially unchanged. But if the thermocouple probe has been in extended service, this estimate of uncertainty is more appropriate.

Considering the effect of the pipe wall temperature on the total uncertainty, the determination can be improved by measuring T_w with a surface temperature thermocouple probe and a separate digital thermometer. With such a system for measuring T_w, the systematic uncertainty for the probe would be

$$b_{\text{probe}} = \tfrac{1}{2}(0.0075)(750°\text{F} - 32°\text{F}) = 2.7°\text{F} \qquad (\text{E.46})$$

With an estimated systematic standard uncertainty of $2°$R for the installation of the surface probe and a systematic standard uncertainty of $1°$R for the digital thermometer, the systematic uncertainty in T_w would be

$$b_{T_w} = [(2.7)^2 + (2)^2 + (1)^2]^{1/2}$$
$$= 3.5°\text{R} \qquad (\text{E.47})$$

These improvements in the overall measurement system yield

$$\begin{array}{cccc} (\epsilon) & (h) & (T_I) & (T_w) \end{array}$$
$$b_{T_g} = [7\ +\ 58\ +\ 25\ +\ 0.5]^{1/2} \qquad (\text{E.48})$$
$$= 9.5°R$$

and
$$T_g = 1520 \pm 19°\text{R} \qquad (\text{E.49})$$

The convective heat transfer coefficient systematic uncertainty of 25% cannot be improved. The effect of this term by itself gives a systematic uncertainty in T_g of $15°$R. Therefore, the only additional improvement in the gas temperature determination would come from calibration of the thermocouple probe and digital thermometer system. With calibration, b_{T_g} could be reduced to less than $0.5°$R, yielding an uncertainty in T_g of

$$T_g = 1519 \pm 16°\text{R} \qquad (\text{E.50})$$

The example above illustrates the sequence of logic used in designing the gas temperature measurement system. The specific systematic uncertainties used and the associated uncertainty of the gas temperature at a 95% level of confidence will depend on the particular application.

REFERENCES

1. Moffat, R. J., "Describing the Uncertainties in Experimental Results," *Experimental Thermal and Fluid Science*, Vol. 1, Jan. 1988, pp. 3–17.

2. Steele, W. G., Maciejewski, P. K., James, C. A., Taylor, R. P., and Coleman, H. W., "Asymmetric Systematic Uncertainties in the Determination of Experimental Uncertainty," *AIAA Journal*, Vol. 34, No. 7, July 1996, pp. 1458–1463.

3. Obert, E. F., *Internal Combustion Engines and Air Pollution*, Harper & Row, New York, 1973.

4. Omega Engineering, *Complete Temperature Measurement Handbook and Encyclopedia*, Omega Engineering, Stamford, CT, 1987, p. T37.

5. Incropera, F. P., and DeWitt, D. P., *Fundamentals of Heat Transfer*, Wiley, New York, 1981.

6. Hottel, H. C., and Sarofim, A. F., *Radiative Transfer*, McGraw-Hill, New York, 1967.

APPENDIX F

DYNAMIC RESPONSE OF INSTRUMENT SYSTEMS

In our discussions in this text, any temporal variations in the measured quantity have been treated as random variations that contribute to the random uncertainty in the measurement. However, it is also necessary for us to consider additional error sources due to the response of instruments to dynamic, or changing, inputs. An instrument may produce an output with both amplitude and phase (time lag) errors when a dynamic input is encountered.

These dynamic response errors are similar to the variable but deterministic bias errors discussed in Appendix E, and they can be very important in the analysis of a timewise, transient experiment. In the following sections we present the fundamentals needed to estimate these amplitude and phase errors.

F-1 GENERAL INSTRUMENT RESPONSE

The traditional way to investigate the dynamic response of an instrument is to consider the differential equation that describes the output. We assume that the instrument response can be modeled using a linear ordinary differential equation with constant coefficients [1]

$$a_n \frac{d^n y}{dt^n} + a_{n-1} \frac{d^{n-1} y}{dt^{n-1}} + \cdots + a_1 \frac{dy}{dt} + a_0 y = bx(t) \tag{F.1}$$

where y is the instrument output, x is the input, and n is the order of the instrument.

Instrument response to three different inputs will be discussed: (1) a step change, (2) a ramp input, and (3) a sinusoidal input. These are illustrated in

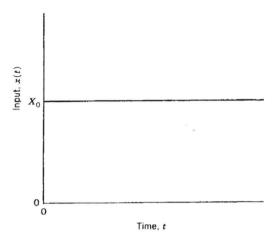

Figure F.1   Step change in input to an instrument.

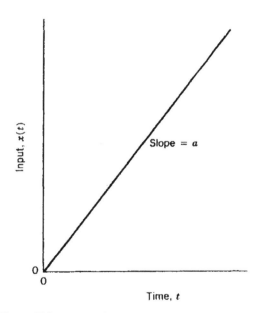

Figure F.2   Ramp change in input to an instrument.

Figures F.1, F.2, and F.3. Mathematically, these inputs are described as follows:

1. Step change $x = 0$ $t < 0$

                    $x = x_0$ $t > 0$
                                                                    (F.2)

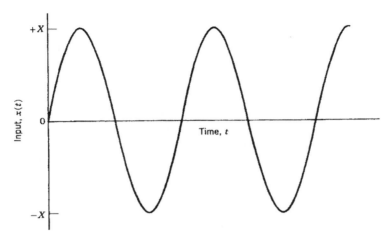

Figure F.3 Sinusoidally varying input to an instrument.

2. Ramp $\qquad\qquad x = 0 \qquad\qquad\qquad\qquad t < 0$

$\qquad\qquad\qquad\qquad x = at \qquad\qquad\qquad\qquad t \geq 0$ $\qquad$ (F.3)

3. Sinusoidal $\qquad x = X\sin(\omega t) \qquad\qquad\quad t > 0$ $\qquad$ (F.4)

The response of zero-, first-, and second-order instruments to these inputs are considered next.

F-2 RESPONSE OF ZERO-ORDER INSTRUMENTS

Since $n = 0$ for a zero-order instrument, Eq. (F.1) reduces to an algebraic equation:

$$y = Kx(t) \qquad\qquad (F.5)$$

where $K \; (= b/a_0)$ is called the *static gain*. Equation (F.5) shows that the output is always proportional to the input, so there is no error in the output due to the dynamic response. Of course, there will be static errors of the types we have discussed previously.

An example of a zero-order instrument is an electrical resistance strain gauge. The input strain ϵ causes the gauge resistance to change by an incremental amount ΔR according to the relationship [2]

$$\Delta R = FR\epsilon \qquad\qquad (F.6)$$

where F is the gauge factor and R is the resistance of the gauge wire in the unstrained condition. Since the instrument itself, the gauge wire, is experiencing the input strain directly, there is no dynamic response error in the output.

F-3 RESPONSE OF FIRST-ORDER INSTRUMENTS

The response equation for first-order instruments is usually written in the form

$$\tau \frac{dy}{dt} + y = Kx \qquad (F.7)$$

where τ $(= a_1/a_0)$ is the time constant and K $(= b/a_0)$ is the static gain. The definition of a first-order instrument is one that has a dynamic response behavior that can be expressed in the form of Eq. (F.7) [3].

A first-order instrument experiences a time delay between its output and a time-varying input. An example is a thermometer or thermocouple that must undergo a heat transfer process for its reading to respond to a changing input temperature.

The response of a first-order instrument to a step change is found by solving Eq. (F.7) using Eq. (F.2) for x and the initial condition that $y = 0$ at $t = 0$. This solution can be expressed in the form

$$\frac{y}{Kx_0} = 1 - e^{-t/\tau} \qquad (F.8)$$

which is plotted in Figure F.4. In one time constant, the response achieves 63.2% of its final value. One must wait for four time constants (4τ) before the response y will be within 2% of the final value.

The response to a ramp input is found by solving Eq. (F.7) using Eq. (F.3) for x and the initial condition that $y = 0$ at $t = 0$. This solution is

$$y = Ka[t - \tau(1 - e^{-t/\tau})] \qquad (F.9)$$

which can also be expressed as

$$y - Kat = -Ka\tau(1 - e^{-t/\tau}) \qquad (F.10)$$

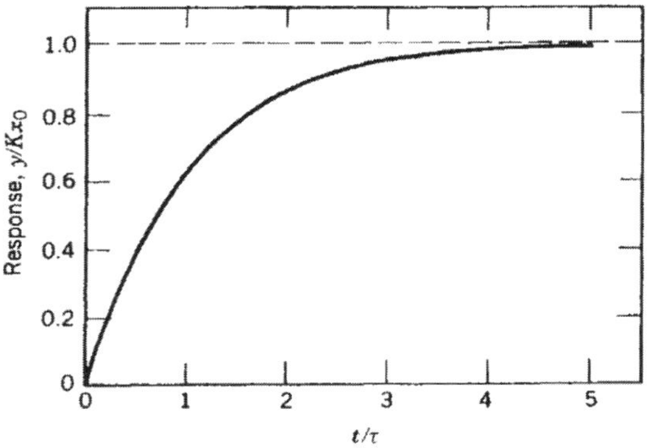

Figure F.4 Response of a first-order instrument to a step change input versus nondimensional time.

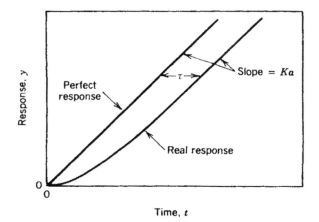

Figure F.5 Response of a first-order instrument to a ramp input versus time.

Equation (F.9) is plotted in Figure F.5. For no dynamic response error, we would obtain $y = Kat$ and the right-hand side (RHS) of Eq. (F.10) would be zero. The two terms on the RHS therefore represent the error in the response. The exponential term $(Ka\tau e^{-t/\tau})$ dies out with time and is called the *transient error*. The other term $(-Ka\tau)$ is constant and proportional to τ. The smaller the time constant is, the smaller this steady-state error will be. The effect of the steady-state error is that the output does not correspond to the input at the current time but to the input τ seconds before.

The response of a first-order instrument to a sinusoidal input is found by solving (Eq. F.7) using Eq. (F.4) for x. This solution is

$$y = Ce^{-t/\tau} + \frac{KX}{\sqrt{1 + \omega^2 \tau^2}} \sin(\omega t + \phi) \tag{F.11}$$

where

$$\phi = \tan^{-1}(-\omega\tau) \tag{F.12}$$

and C is the arbitrary constant of integration. The exponential term in Eq. (F.11) is the transient error that dies out in a few time constants. The second term on the RHS of Eq. (F.11) is the steady sinusoidal response of the instrument.

By comparing the steady response with the input [Eq. (F.4)], we see that the response has an amplitude error proportional to the amplitude coefficient $(1/\sqrt{1 + \omega^2 \tau^2})$ and a phase error ϕ. These errors are shown in Figures F.6 and F.7. Each of these errors varies with the product of the time constant τ and the frequency of the input signal ω. As $\omega\tau$ increases, the amplitude coefficient decreases and the deviation from a perfect response becomes greater and greater, as seen in Figure F.6. A similar behavior is observed in Figure F.7 for the phase error, which asymptotically approaches $-90°$ as $\omega\tau$ increases.

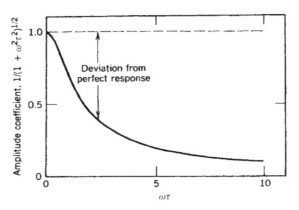

Figure F.6   Amplitude response of a first-order instrument to a sinusoidal input of frequency ω.

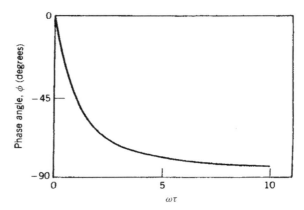

Figure F.7   Phase error in the response of a first-order instrument to a sinusoidal input of frequency ω.

F-4 RESPONSE OF SECOND-ORDER INSTRUMENTS

With $n = 2$ in Eq. (F.1), the response of a second-order instrument becomes

$$a_2 \frac{d^2 y}{dt^2} + a_1 \frac{dy}{dt} + a_0 y = b x(t) \tag{F.13}$$

If this expression is divided through by a_2, it can be written in the form

$$\frac{d^2 y}{dt^2} + 2\zeta \omega_n \frac{dy}{dt} + \omega_n^2 y = K \omega_n^2 x(t) \tag{F.14}$$

where $K(= b/a_0)$ is again the static gain, $\zeta(= a_1/2\sqrt{a_0a_2})$ is the damping factor, and $\omega_n(= \sqrt{a_0/a_2})$ is the natural frequency. The definition of a second-order instrument is one that has a dynamic response behavior that can be expressed in the form of Eq. (F.14) [3]. Instruments that exhibit a spring–mass type of behavior are second order. Examples are galvanometers, accelerometers, diaphragm-type pressure transducers, and U-tube manometers [1].

The nature of the solutions to Eq. (F.14) is determined by the value of the damping constant ζ. For $\zeta < 1$, the system is said to be *underdamped* and the solution is oscillatory. For $\zeta = 1$, the system is *critically damped*, and for $\zeta > 1$ the system is said to be *overdamped*.

The second-order instrument response to a step change is found by solving Eq. (F.14) using Eq. (F.2) for x and the initial conditions that $y = y' = 0$ at $t = 0$. The solution depends on the value of ζ and is given by:

$\zeta > 1$:

$$y = Kx_0 \left\{ 1 - e^{-\zeta\omega_n t} \left[\cosh(\omega_n t \sqrt{\zeta^2 - 1}) + \frac{\zeta}{\sqrt{\zeta^2 - 1}} \sinh(\omega_n t \sqrt{\zeta^2 - 1}) \right] \right\}$$

(F.15)

$\zeta = 1$:

$$y = Kx_0[1 - e^{-\omega_n t}(1 + \omega_n t)]$$

(F.16)

$\zeta < 1$:

$$y = Kx_0 \left\{ 1 - e^{-\zeta\omega_n t} \left[\frac{1}{\sqrt{1 - \zeta^2}} \sin(\omega_n t \sqrt{1 - \zeta^2} + \phi) \right] \right\}$$

(F.17)

where

$$\phi = \sin^{-1}(\sqrt{1 - \zeta^2})$$

(F.18)

This response is shown in Figure F.8. Note that $1/\zeta\omega_n$ is now the time constant. The larger $\zeta\omega_n$ is, the more quickly the response approaches the steady-state value. The form of the approach to the steady-state value is determined by ζ. For $\zeta < 1$, the response overshoots, then oscillates about the final value while being damped.

Most instruments are designed with damping factors of about 0.7. The reason for this can be seen in Figure F.8. If an overshoot of 5% is allowed, a damping factor $\zeta \simeq 0.7$ will result in a response that is within 5% of the steady-state value in about half the time required by an instrument with $\zeta = 1$ [1]. Note that the steady-state solution for all values of $\zeta > 0$ gives Kx_0.

The response to a ramp input also contains a transient and steady-state portion. The steady-state solution is

$$y = Ka \left(t - \frac{2\zeta}{\omega_n} \right)$$

(F.19)

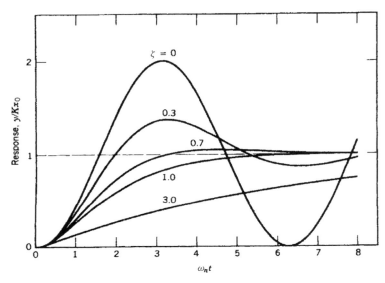

Figure F.8   Response of a second-order instrument to a step change input for various damping factors.

The response lags behind the input by a time equal to $2\zeta/\omega_n$. High values of ω_n and/or low values of ζ reduce this lag in the steady-state response.

The response of a second-order instrument to a sinusoidal input is (at steady state) given by

$$y = \frac{KX}{[(1 - \omega^2/\omega_n^2)^2 + (2\zeta\omega/\omega_n)^2]^{1/2}} \sin(\omega t + \phi) \qquad (F.20)$$

where

$$\phi = \tan^{-1}\left(-\frac{2\zeta\omega/\omega_n}{1 - \omega^2/\omega_n^2}\right) \qquad (F.21)$$

As in the first-order system, the response contains both an amplitude error proportional to an amplitude coefficient and a phase error. These errors are shown in Figures F.9 and F.10.

From Figure F.9 we see that for no damping ($\zeta = 0$) the amplitude of the response approaches infinity as the input signal frequency ω approaches the instrument natural frequency ω_n. In general, this maximum amplitude, or resonance, will occur at

$$\omega = \omega_n\sqrt{1 - 2\zeta^2} \qquad (F.22)$$

Note that for nonzero damping (which is always the case physically) the amplitude at this resonant frequency will be finite. Also note that for $\zeta \simeq 0.6$ to 0.7 and

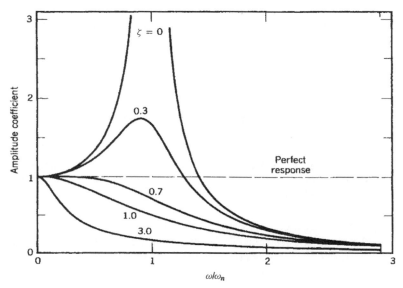

Figure F.9 Amplitude response of a second-order instrument to a sinusoidal input of frequency ω.

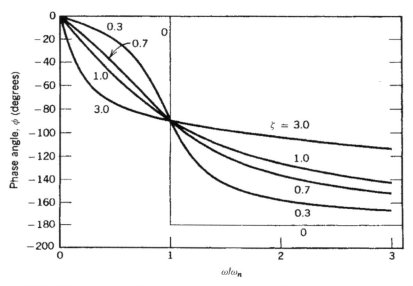

Figure F.10 Phase error in the response of a second-order instrument to a sinusoidal input of frequency ω.

$\omega/\omega_n < 1$, the amplitude error is minimized and the phase error is about linear, which is desirable because this produces minimum distortion of the input signal [1].

F-5 SUMMARY

The dynamic response of zero-, first-, and second-order instruments has been presented for step, ramp, and sinusoidal inputs. In the case of a zero-order instrument, it was found that there is no dynamic response error. For first- and second-order instruments, there are time delays for step and ramp inputs and amplitude and phase errors for sinusoidal inputs. By choosing or designing instruments with appropriate values of time constant and natural frequency, the effects of these errors can be minimized.

In a complete measurement system, different instruments will usually be connected in the form of a transducer, a signal conditioning device, and a readout. An example might be a thermocouple (a first-order instrument) connected to an analog voltmeter (a second-order instrument). In such cases the dynamic output of the first instrument can be determined and this value can be used as the input to the second instrument. The dynamic response of the second instrument to this input is then determined to obtain the dynamic response of the system. In most cases the significant dynamic response error effect will occur with only one of the instruments, usually the transducer.

REFERENCES

1. Schenck, H., *Theories of Engineering Experimentation*, 3rd ed., McGraw-Hill, New York, 1979.
2. Holman, J. P., *Experimental Methods for Engineers*, 4th ed., McGraw-Hill, New York, 1984.
3. Doebelin, E. O., *Measurement Systems Application and Design*, 3rd ed., McGraw-Hill, New York, 1983.

INDEX